Environmental Impact Assessment

Ultimate Tools, Techniques and Technology to Natural Resources Management

Dr. Anthonysamy David

Table of Contents

Acknowledgement 1
Disclaimer 4
INTRODUCTION 6
Chapter 1 25
Introduction to Environmental Impact Assessment 25
Concept and Purpose of EIA 27
Historical Background and Evolution 33
Importance in Sustainable Project Planning 36
Stakeholder Involvement 41
Case Studies Showcasing EIA Significance 46
Summary and Reflections 53
Reference List 54
Chapter 2 58
Key Elements of the EIA Process 58
Screening 59
Scoping 64
Baseline Data Collection 68
Impact Analysis 73
Process Challenges 77
Summary and Reflections 83

Chapter 3 85
Technological Tools in EIA 85
- Geographic Information Systems (GIS) 86
- Remote Sensing Techniques 90
- Environmental Modelling 95
- Data Analytics in Environmental Assessments 100
- Big Data Utilisation: 101
- Real-Time Data Analysis: 102
- Machine Learning Applications: 103
- Risk Assessment Enhancement: 104
- Stakeholder Engagement Tools 105
- Insights and Implications 109

Reference List 111

Chapter 4 114
Mitigation Measures and Strategies 114
- Design Adjustments for Mitigation 115
- Compensatory Measures 120
- Innovative Technological Solutions 126
- Long-Term Monitoring Plans 130
- Integration of Sustainable Architecture 135
  - Promoting Resource Conservation and Energy Efficiency 136
  - Encouraging Eco-Friendly Construction Techniques 137

Reducing Waste During the Project Lifecycle 138
Aligning with Circular Economy Principles 139
Promoting Resource Conservation and Energy Efficiency in the Design Phase 140
Encouraging Adoption of Eco-Friendly Construction Techniques Among Practitioners 141
Contributing to the Reduction of Waste During the Project's Lifecycle and Supporting Sustainability 142
Aligning with Circular Economy Principles That Focus on Resource Efficiency 143
Final Thoughts 144
Reference List 145
Chapter 5 149
Public Consultation and Stakeholder Engagement 149
Methods of Public Engagement 150
Examples of Community Involvement 155
Challenges in Stakeholder Consultations 160

Benefits of Inclusive Decision-Making 164
Effective Facilitation of Public Engagement 171
Final Thoughts 176
Chaper 6 179
Environmental Components in EIA 179
Air Quality Assessment 180
Water Resources Evaluation 186
Soil Health Analysis 193
Flora and Fauna Impacts 198
Biodiversity Assessment Techniques 198
Impacts of Habitat Alteration 200
Conservation Strategies 201
Ecological Value of Biodiversity 202
Integrated Impact Mitigation Strategies 203
Concluding Thoughts 208
Chapter 7 210
Social and Economic Considerations 210
Health Impact Assessments 211
Safety and Risk Evaluations 217
Analysis of Livelihoods 222
Community Cohesion Impacts 227
Recommendations and Strategies 232
Insights and Implications 236
Chapter 8 238

Principles of Sustainable Development 238
- The Concept of Sustainable Development 240
- Precautionary Principle 246
- Polluter Pays Principle 251
- Intergenerational Equity 256
- Examples of Sustainable Development in Practice 262
- Summary and Reflections 265
- Reference List 266

Chapter 9 270
Sector-Specific EIA Approaches 270
- Mining Projects and Their Impacts 271
- Urban Development and Planning 277
- Coastal Habitat Protection 282
- Sustainable Business Practices in Various Sectors 288
- Water Contamination Issues 294
- Insights and Implications 299
- Reference List 300

Chapter 10 304
Future Trends and Innovations in EIA 304
- Advancements in Remote Sensing and GIS 306
- Impact of AI and Machine Learning 311
- Role of Citizen Science 317

Policy and Regulatory Changes on the Horizon 323
Technological Innovations Improving EIA Efficiency 331
Final Thoughts 335
Reference List 336
Conclusion 340

Acknowledgement

Environmental Impact Assessment: Ultimate Tools, Techniques and Technology to Natural Resources Management is a comprehensive guide to the methodologies, tools, and technologies involved in environmental impact assessments, as well as their critical role in natural resource management. The book aims to provide a thorough understanding of the Environmental Impact Assessment (EIA) process, explore various tools, techniques, and technologies used in conducting EIAs, and engage readers in a discourse on the critical relationship between environmental assessment and natural resource management.

The target audience for this book includes environmental professionals, policymakers, academics, and students interested in sustainability, environmental protection, and resource management. The book offers an integrative approach that combines theoretical foundations with practical tools, technologies, and case studies, tailored to address contemporary challenges in environmental assessments while encouraging stakeholder engagement in the decision-making process.

The EIA process helps to understand how proposed projects may affect the environment by identifying, predicting, and evaluating both the positive and negative impacts of a project before it begins. It assists decision-makers in understanding the potential trade-offs involved, and public involvement is essential for transparency and community engagement. Engaging stakeholders in meaningful public consultations can lead to better project designs that take community feedback into account.

One of the significant roles of an EIA is to develop strategies that mitigate adverse environmental impacts, finding ways to avoid or reduce harmful effects on the environment. For instance, if the EIA anticipates a project's disruption of a local habitat, it may recommend alternative project locations or design modifications to mitigate the impact.

In conclusion, The Ultimate Guide to Environmental Impact Assessment and Resource Management Tools provides a comprehensive overview of the methodologies, tools, and technologies involved in environmental evaluation and resource management. It offers valuable insights for practitioners, policymakers, and students interested in sustainable development and environmental protection.

I would like to thank the Archbishop of Patna, Sebastian Kallupura; Vicar General Fr. James George; the treasurer, Fr. Amal Raj; and all my dear priest friends in the Archdiocese of Patna, who always inspire me to undertake such projects. Last but not least, I would like to thank the provincial of the Pune Jesuit Province, Fr. Agnello Macarenhas, and the sociologist, Fr. Stan Fernandes, along with all others who are with me in this endeavor, providing constant support and guidance.

Disclaimer

This document describes work undertaken as part of a programme of study at the Xavier Institute of Natural Resource Management, affiliated with Savitribai Phule Pune University. All views and opinions expressed therein remain the sole responsibility of the author and do not necessarily represent those of the faculty. It is the property of Dr. Anthony Samy.

Dr. Anthonysamy David
Xavier Institute of Natural Resources Management
Ahmednagar, Maharashtra, 4141001
danthonysamy@gmail.com

Environmental Impact Assessment: Ultimate Tools, Techniques and Technology to Natural Resource Management

First Edition: July 2024

INTRODUCTION

Imagine standing at the edge of a pristine forest, knowing that a new highway could soon slice through its heart, altering the landscape forever. The Environmental Impact Assessment (EIA) process is designed to foresee these changes before they happen, advocating for our planet. This scenario encapsulates the essence of what this book seeks to explore—a detailed guide on how EIAs can influence development while safeguarding our environmental heritage.

In a world grappling with climate change, deforestation, and rapid urbanisation, understanding how projects influence our environment is not just a matter of policy but a necessity for survival and sustainability. Today, more than ever, there is an urgent need for meticulous evaluation of how our actions impact the natural world. For environmental professionals and practitioners, this book offers valuable methodologies and practical tools for conducting effective assessments, enhancing their expertise in the field. Policymakers and regulators invested in sustainable development will find this work beneficial in crafting more effective environmental policies and guidelines. Students and academics pursuing environmental science, natural resource management, and sustainability programmes will gain foundational knowledge that intricately weaves theory with practical application.

The EIA process serves as a vital roadmap, guiding decision-makers to identify potential environmental impacts through stages like screening, scoping, and public consultation, ensuring that proposed projects harmonise with nature rather than disrupt it. Understanding this process is crucial for all stakeholders involved in environmental decision-making. It equips them with the ability to foresee adverse effects and implement strategies to mitigate harm, thus promoting development that aligns with sustainable principles.

EIAs are significant in today's context for various reasons. First, they provide a structured framework for evaluating the environmental ramifications of projects before they commence, which helps in avoiding or minimising environmental damage. This approach is especially pertinent given the current global challenges, such as climate change, wherein every project has the potential to either mitigate or exacerbate the problem. Second, EIAs foster transparency and public involvement, essential components for democratic decision-making and fostering community trust.

Have you ever considered the lasting legacy of a single construction project in your neighbourhood? From the air we breathe to the wildlife that shares our spaces, the outcomes of these developments hinge on comprehensive environmental assessments. Personalising the importance of EIA can lead to greater awareness and engagement from the public. When individuals understand the direct impacts on their lives and communities, support for sustainable practices strengthens.

To appreciate the breadth and depth of EIAs, one must first recognise their multi-faceted nature. It is both a scientific discipline and a public engagement process. Scientifically, it involves rigorous data collection, analysis, and prediction of environmental impacts using various methodologies and tools. Publicly, it mandates consultation and participation, ensuring that the concerns of affected communities are heard and addressed. This dual aspect underscores the importance of EIAs in fostering informed and inclusive decision-making.

Conducting an EIA requires a systematic and iterative approach. It begins with screening to determine whether a project requires a detailed assessment. This step ensures resources are directed towards projects with significant potential impacts. Next comes scoping, which identifies key issues and sets the boundaries for the assessment. This phase includes collecting baseline data, predicting impacts, and evaluating alternatives to prevent or reduce negative effects. Public consultation follows, offering stakeholders the opportunity to voice their opinions and contribute to the decision-making process. Finally, the findings and recommendations are documented in an EIA report, which guides regulatory bodies and other stakeholders in approving or modifying the proposed project.

The significance of including public opinion cannot be overstated. Engaging communities early in the process often leads to better project design and execution, as local insights and traditional knowledge can offer invaluable perspectives. Furthermore, it builds trust and reduces conflicts that might arise from misunderstandings or lack of information.

Implementing recommendations from an EIA also involves monitoring and managing identified impacts. This phase ensures that mitigation measures are effectively carried out and unforeseen issues are promptly addressed. Continuous monitoring helps maintain accountability and provides data for future EIAs, creating a cycle of improvement and refinement.

Real-world examples illustrate the value of EIAs. Consider a proposed dam project in a developing country. Without an EIA, the project might proceed, leading to potential flooding of vast areas, displacement of communities, destruction of habitats, and alteration of water quality and flow regimes. An EIA would predict these impacts, propose alternative designs, suggest mitigation measures such as building fish ladders, and involve local communities in decision-making, thereby balancing development needs with environmental protection.

In urban settings, EIAs can dictate how new infrastructure projects like highways, airports, or industrial zones coexist with existing natural and human environments. For instance, an EIA might reveal that constructing a new road through a wetland area would degrade water quality and disrupt species habitats. It can recommend rerouting the road or incorporating wildlife corridors to preserve ecological integrity while meeting transportation needs.

By providing a structured methodology, EIAs empower environmental professionals to anticipate and manage the ecological consequences of development projects. Policymakers gain insight into the long-term sustainability implications of their decisions, fostering a more holistic approach to economic growth. For students and academics, this book bridges the gap between theoretical concepts and practical applications, enriching their academic journey and preparing them for future challenges in environmental stewardship.

This introduction sets the stage for a comprehensive exploration of the EIA process. It highlights the critical role EIAs play in sustainable development and emphasises the collective responsibility of professionals, policymakers, and the public in nurturing our planet for future generations. As we delve deeper into the chapters ahead, we will uncover the intricate details of conducting EIAs, the methodologies employed, and case studies that exemplify their successes and failures. By the end of this book, readers will be equipped with the knowledge and tools to make informed decisions that prioritise environmental health alongside human progress, ensuring a balanced and sustainable future.

The Environmental Impact Assessment (EIA) is a crucial process that helps identify, predict, and evaluate the environmental impacts of proposed projects before they begin. It aims to ensure that projects are environmentally friendly and sustainable for the future. The EIA process involves several key elements, including screening, scoping, baseline data collection, impact analysis, mitigation measures, public consultation, reporting, review and decision-making, and monitoring and compliance.

Screening determines whether an EIA is necessary for a specific project and the level of assessment required. Scoping highlights key environmental issues and potential impacts, while baseline data collection gathers detailed information about the surrounding ecosystem. Impact analysis uses various methods, such as environmental models, expert judgement, and comparative analysis, to predict and evaluate potential environmental impacts. Mitigation measures are developed to reduce or eliminate adverse effects on the environment, such as by redesigning project plans, using new technologies, or implementing compensatory measures.

Public consultation is essential in the EIA process, as it helps incorporate stakeholders' opinions and feedback into the assessment. Public meetings, surveys, and focus groups can be used to gather input from the community, raising awareness and understanding among local residents.

After the assessment is complete, the findings are documented in an Environmental Impact Statement (EIS), which serves as a key document for decision-makers before making final decisions about project approval. The review and decision-making stage involves assessing the EIS and making a final decision regarding the project. Monitoring and compliance are necessary to ensure effective implementation of mitigation measures, such as regular site inspections and environmental monitoring programmes. By following these steps, the EIA process ensures responsible and respectful project conduct.

Environmental components are crucial in understanding the impact of projects on our surroundings. Air quality, water resources, soil health, flora and fauna, noise pollution, cultural heritage, social components, economic components, and conditions and principles guide a sustainable environmental assessment. Runoff and pollution affect water resources, while emissions and pollutants have an impact on air quality. Habitat loss and species displacement affect flora and fauna, while erosion and contamination affect soil health. Noise pollution is another issue, with high levels of noise leading to stress and sleep disturbances in local communities. Projects that involve the destruction of historical sites or artefacts also have an impact on cultural heritage.

Social components involve evaluating potential health impacts on local communities, safety, livelihoods, and social cohesion. Projects can introduce pollutants and stressors that can harm residents' health, and safety measures should be developed to protect local populations. Developments impact livelihoods, and it is important to take social cohesion into account because projects can disrupt long-standing relationships. Economic components involve comparing economic benefits with environmental costs, assessing economic sustainability, and examining the effects on local economies.

Conditions and principles guide a sustainable environmental assessment, emphasising sustainable development, the precautionary principle, the polluter pays principle, and public participation. Sustainable development emphasises meeting present needs without compromising future generations, while the precautionary principle promotes preventive action. The polluter pays principle ensures responsible parties bear the costs associated with pollution, motivating companies to adopt cleaner practices and technologies. Public participation creates transparency and builds trust, leading to more informed and accepted outcomes.

Environmental Impact Assessments (EIA) are crucial for assessing the long-term impacts of various projects and developments. Strategic Environmental Assessment (SEA) focusses on policies, plans, and programmes, while project-level EIA evaluates specific projects or developments. Cumulative impact assessments examine the combined impacts of multiple projects, particularly in regions where developments occur simultaneously or in close proximity.

A multidisciplinary approach is beneficial, involving expertise from various fields such as ecology, sociology, economics, and engineering. Transparency and accountability are essential for a successful EIA process, fostering trust and community involvement. Capacity building is also important, enabling institutions and stakeholders to participate effectively in and conduct EIAs.

Equitable use of resources is a principle that guides resource sharing and management. It revolves around two main concepts: resource allocation and intergenerational equity. Resource allocation involves the fair distribution of natural resources, such as water, land, and minerals, without favouring one group over another. Balancing these needs is critical for equitable resource use.

Intergenerational equity focusses on protecting the interests of future generations, such as forests. To ensure sustainability, practices that support both current needs and those of future generations should be adopted. Investing in renewable resources, educating people about sustainable practices, and incorporating policies that promote sustainability are essential.

Transparency in resource allocation ensures everyone knows who is using what and why, allowing individuals to express concerns. Establishing forums for discussion and regular meetings can facilitate dialogue about resource use in communities or organisations. Technology can also assist in achieving better resource allocation, such as monitoring water usage in agriculture. By recognising our responsibility as individuals and as part of a larger community, we can work towards a healthier planet for ourselves and future generations.

Sustainable business practices are becoming increasingly important in today's world, with many companies realising their responsibility to contribute to environmental well-being. Corporate responsibility encourages businesses to adopt environmentally friendly practices, such as investing in cleaner technologies, sourcing materials from sustainable suppliers, and reducing their environmental footprint through recycling programmes. Green governance is another crucial concept in the realm of sustainable business, emphasising the need for businesses to follow guidelines that protect the environment while encouraging responsible resource management.

To effectively implement green governance, businesses should conduct audits of their current practices, identify areas for improvement, and train employees on sustainable practices. Collaborating with local communities and partnering with community organisations can also help promote green initiatives.

Urban development projects involve various techniques and tools to ensure environmental changes are managed well. Remote sensing and Geographic Information Systems (GIS) can help assess the impacts of urban development proposals, especially in rapidly growing cities. Land use change detection, air quality modelling, socio-economic impact assessment, and environmental impact assessment (EIA) are essential tools for assessing the impacts of urban development projects.

EIA helps identify areas where green spaces can be preserved or created, mitigating air pollution and planning for infrastructure. By working with residents to find solutions that meet the development's needs without forcing people to leave their homes, businesses can become part of the neighbourhood rather than an external force disrupting it.

In conclusion, sustainable business practices, corporate responsibility, and urban development projects all play a crucial role in promoting environmental sustainability and ensuring the well-being of the environment. By implementing these practices, businesses can contribute positively to the environment and the overall well-being of their communities.

Urban development projects require collaboration among various stakeholders, including city planners, environmental scientists, community members, and government officials. This collaboration is key to making informed decisions that meet developmental goals and community needs. By integrating green spaces, improving air quality measures, and thoughtfully planning infrastructure, urban development can lead to sustainable growth and create a healthy environment for residents.

Urban development can enhance the quality of life for current and future residents by using remote sensing and GIS to inform and guide the planning process. Mining projects, such as mining, can lead to negative environmental impacts like deforestation and land degradation. To address these challenges, mining companies have adopted remote sensing techniques, using satellite images and aerial photography to monitor changes in land caused by mining activities.

Geographic Information Systems (GIS) are used to analyse water quality data, allowing companies to make informed decisions about protecting natural resources and ensuring community safety. These analyses also help predict future impacts of mining on water sources, which are essential for maintaining ecosystem health and providing safe drinking water.

The Environmental Impact Assessment (EIA) evaluates the potential environmental effects of a proposed project, highlighting vulnerabilities created by mining activities and developing strategies to mitigate them. Following the EIA, the company implemented more sustainable practices, including reforestation, water treatment plans, and community involvement.

These efforts have led to positive outcomes both environmentally and socially, reducing deforestation, improving water quality, and generating goodwill within the communities they operate in. By taking responsibility for their impact on the environment and proactively addressing issues, the mining company demonstrated that it is possible to balance industrial activity with environmental preservation.

The Environmental Impact Assessment (EIA) is a crucial tool for mining companies to monitor and analyse the impacts of their activities. It highlights the importance of engaging stakeholders and committing to sustainable practices to protect the environment and maintain positive relationships with communities. Coastal habitats, such as beaches, wetlands, and mangroves, are vital for maintaining healthy marine environments and supporting biodiversity. To assess the impacts of construction, GIS is used to analyse the spatial distribution of habitats and identify sensitive habitats.

Mitigation measures, such as creating buffer zones, are crucial to lessening the impact of construction activities on the environment. These zones help maintain natural conditions for sensitive habitats to thrive. Implementing buffer zones, such as vegetated areas or designated zones, ensures marine life has a chance to survive despite nearby construction activities.

Conservation areas are also essential for protecting marine habitats. They provide safe spaces for endangered species and act as nurseries for juvenile fish. These areas can enhance biodiversity, serve as research sites for scientists studying marine ecosystems, track species population changes, and attract eco-tourism, bringing economic benefits to local communities.

In conclusion, the EIA serves as a vital tool for ensuring the well-being of marine ecosystems in construction projects. By prioritising the protection of critical marine habitats, companies can continue to safeguard their precious coastal habitats for future generations.

Chapter 1

Introduction to Environmental Impact Assessment

Environmental Impact Assessment (EIA) is a systematic method to evaluate the environmental consequences of proposed projects. It ensures that potential impacts on the natural environment are considered before any project begins. By identifying and analysing these impacts, EIA helps in mitigating adverse effects and promoting sustainable development practices. This chapter serves as an introduction to the fundamentals and significance of Environmental Impact Assessments, laying the groundwork for understanding their role in sustainable project planning and execution.

The chapter delves into the concept and purpose of EIA, highlighting its importance in predicting both positive and negative environmental impacts. It also explores how EIAs function as essential decision-making tools for various stakeholders, such as government agencies, project proponents, local communities, and environmental organisations. Additionally, the chapter underscores the collaborative nature of the EIA process, involving experts from multiple disciplines to ensure comprehensive assessments. It outlines the structured steps involved in conducting an EIA, like screening, scoping, data collection, impact prediction, public consultation, and formulation of mitigation measures. Through this detailed exploration, readers will gain a thorough understanding of how EIAs contribute to more environmentally sound and socially responsible project outcomes.

Concept and Purpose of EIA

Environmental Impact Assessment (EIA) is a systematic process designed to evaluate the environmental consequences of proposed projects. It involves a comprehensive analysis to identify, predict, and interpret the potential impacts that a project may have on the natural environment before it begins. The primary goal of EIA is to ensure that decision-makers consider the potential environmental effects to mitigate adverse impacts and promote sustainable development.

EIA is crucial because it helps identify potential environmental impacts before project initiation. This proactive approach allows for the identification of both positive and negative effects on the environment, human health, and socioeconomic conditions. Early identification of impacts ensures that appropriate mitigation measures can be planned and implemented to avoid or minimise negative consequences. By assessing potential impacts beforehand, EIA aids in crafting more environmentally sound and socially responsible projects.

Furthermore, EIA serves as an essential decision-making tool for stakeholders. When a project is proposed, various stakeholders—including government agencies, project proponents, local communities, and environmental organizations—need reliable information to make informed decisions. EIA provides a structured framework that systematically evaluates the possible outcomes of a project, enabling stakeholders to weigh environmental considerations alongside economic and social factors. This holistic approach facilitates balanced decision-making, ensuring that developmental objectives do not come at the expense of environmental integrity.

The EIA process is inherently collaborative, requiring the involvement of various experts during different assessment stages. Environmental scientists, ecologists, economists, sociologists, and other specialists contribute their knowledge and expertise to create a comprehensive evaluation of the project's potential impacts. This interdisciplinary collaboration enhances the robustness of the assessment by incorporating diverse perspectives and addressing multiple dimensions of the environment. For instance, ecologists might study the impact on local flora and fauna, while sociologists could assess the project's implications for nearby communities.

The definition of EIA as a systematic process underscores its structured approach to evaluating environmental consequences. Systematic evaluation means that the process is organised, methodical, and follows a predefined sequence of steps. These steps often include screening to determine if an EIA is necessary, scoping to identify significant issues to be evaluated, data collection and analysis, impact prediction and evaluation, public consultation, and the formulation of mitigation measures. Each step is crucial to ensuring a thorough and unbiased assessment, which ultimately contributes to more sustainable project outcomes.

Identification of potential impacts before project initiation is another critical aspect of EIA. By conducting this assessment early in the planning phase, project proponents can incorporate environmental considerations into the project design, thereby avoiding irreversible damage. For example, if an EIA reveals that a proposed construction project will disrupt a sensitive wetland area, alternative site locations can be explored or design modifications can be made to minimise the disturbance. Such preemptive actions not only protect the environment but also save time and resources by preventing delays and conflicts during later stages of the project.

EIA also plays a vital role as a decision-making tool for stakeholders. Decision-makers rely on EIAs to provide objective and scientifically backed information about the potential environmental impacts of a project. This information is crucial for balancing developmental needs with environmental protection. For instance, policymakers may use the findings of an EIA to regulate the permissible levels of emissions from an industrial facility to protect air quality. Similarly, financial institutions might use EIA reports to assess the environmental risks associated with financing a project, ensuring that investments align with sustainability principles.

Collaboration among various experts during assessment stages further strengthens the EIA process. The complexity of environmental systems necessitates input from multiple disciplines to capture the full range of potential impacts. Specialists in soil science, hydrology, air quality, wildlife biology, and community health, among others, contribute their expertise to developing a comprehensive understanding of how a project might interact with different environmental components. This multidisciplinary approach ensures that the assessment is thorough and considers all relevant factors.

Guidelines are essential when discussing the key components of EIA, stakeholder involvement, and legal and regulatory frameworks. Guidelines ensure consistency and standardisation in the assessment process, helping maintain high-quality evaluations across different projects and jurisdictions.

First, let's explore the key components of EIA. These typically include screening, scoping, baseline studies, impact prediction, mitigation measures, reporting, and monitoring. Screening determines whether an EIA is required for a particular project. Scoping identifies the key environmental issues and concerns that need to be addressed. Baseline studies establish the existing environmental conditions against which changes can be measured. Impact prediction forecasts the likely environmental changes resulting from the project. Mitigation measures propose actions to prevent, reduce, or compensate for adverse impacts. Reporting involves documenting the findings and recommendations in an Environmental Impact Statement (EIS). Finally, monitoring ensures compliance with environmental management plans and verifies the accuracy of impact predictions.

Stakeholder involvement is another crucial component of EIA. Engaging stakeholders, including local communities, interest groups, and the general public, is fundamental to the credibility and effectiveness of the assessment process. Stakeholders provide valuable insights, raise concerns, and contribute local knowledge, leading to more robust and accepted outcomes. Effective stakeholder engagement involves transparent communication, inclusive participation, and consideration of feedback throughout the EIA process.

Legal and regulatory frameworks govern the conduct of EIA, ensuring that assessments meet established standards and criteria. Different countries have varying regulations and guidelines, but common elements include requirements for public participation, transparency, and detailed documentation of the EIA process. Adherence to these frameworks ensures accountability and consistency in environmental assessments.

Historical Background and Evolution

Environmental Impact Assessment (EIA) has undergone significant evolution since its inception, reflecting broader changes in environmental policy and practice. The 1960s were a time of rising environmental consciousness and the adoption of ground-breaking environmental legislation, which is when EIA first appeared. One pivotal milestone was the National Environmental Policy Act (NEPA) of 1969 in the United States, which mandated federal agencies to consider environmental consequences before undertaking any major project. This legislation laid the groundwork for EIA by formalising a process that evaluates potential environmental impacts systematically.

Key legislation and global agreements have played crucial roles in shaping EIA practices worldwide. Following NEPA, many countries adopted similar frameworks, leading to the creation of various international conventions and protocols. For instance, the Espoo Convention on Environmental Impact Assessment in a Transboundary Context, adopted in 1991, emphasised the need for cross-border environmental assessments. Such agreements have made EIA a standard requirement for large-scale projects, ensuring that environmental, social, and economic impacts are considered globally (RJ, 2023).

The transformation from qualitative methods to more structured frameworks marks a significant phase in the evolution of EIA. Initial EIAs were often descriptive and lacked standardised procedures, making comparisons difficult. Over time, however, these methods evolved into robust, quantitative frameworks that provide clear guidelines for assessment processes. The need for greater consistency and objectivity in assessing environmental impacts was what sparked this shift. Frameworks like the Strategic Environmental Assessment (SEA) emerged, allowing for comprehensive analysis at various planning levels—from individual projects to broader policy initiatives (Zhao et al., 2023).

Technological advancements have further revolutionised EIA processes. Early assessments relied heavily on manual data collection and fieldwork, which could be time-consuming and prone to errors. Today, advanced technologies such as Geographic Information Systems (GIS), remote sensing, and computer modelling have significantly enhanced the accuracy and efficiency of EIAs. GIS, for example, allows for the creation of detailed spatial databases and maps, enabling analysts to visualise project impacts with precision. Remote sensing provides up-to-date information about environmental conditions, while computer models can simulate various scenarios to predict potential outcomes. These technologies have not only improved the quality of EIAs but also facilitated better decision-making by providing stakeholders with comprehensive, reliable data.

Importance in Sustainable Project Planning

Environmental Impact Assessments (EIAs) are essential tools in promoting sustainable development and ensuring that environmental considerations are integrated into project planning and execution. The goal of this subpoint is to highlight how EIAs contribute to achieving sustainable project outcomes, reinforcing their necessity in the planning processes. This will be looked into by looking at a number of related ideas, such as how to use environmental factors to make better decisions, how to make changes to projects that make them more sustainable, how good EIAs can improve community satisfaction, and how they can help protect biodiversity and ecosystem stability.

Integration of environmental considerations for informed decision-making is a fundamental aspect of EIAs. By systematically evaluating potential environmental impacts, decision-makers can better understand the consequences of proposed actions and make more informed choices. EIAs typically involve conducting baseline studies, which provide a comprehensive understanding of the current environmental conditions before any development begins. These studies help identify sensitive areas, such as habitats of endangered species or regions with high biodiversity, which need protection. Armed with this information, planners can consider alternative project designs or locations that minimise environmental disruption. For example, an EIA might reveal that a proposed industrial facility could harm a nearby wetland. As a result, the project might be relocated or redesigned to include measures that protect the wetland, demonstrating how EIAs facilitate better decision-making by incorporating environmental considerations early in the planning process.

Guidance for project modifications that enhance sustainability is another critical function of EIAs. When potential adverse impacts are identified during the assessment process, EIAs offer strategies for mitigating these effects. These mitigation measures can range from altering construction methods to using more sustainable materials. For instance, if an EIA identifies that a construction project could lead to significant soil erosion, recommendations might include implementing erosion control techniques such as planting vegetation or building retaining walls. Similarly, if a project threatens local water quality, the EIA might suggest installing advanced wastewater treatment systems. By providing concrete recommendations for project modifications, EIAs help ensure that developments proceed in an environmentally responsible manner, ultimately contributing to more sustainable outcomes. Additionally, EIAs often outline monitoring requirements to track the effectiveness of mitigation measures over time, enabling ongoing adjustments as needed to maintain sustainability.

The correlation between effective EIAs and community satisfaction cannot be overstated. Projects that undergo thorough EIAs tend to foster greater trust and acceptance among local communities. This is because EIAs involve public participation, allowing stakeholders to voice their concerns and contribute to the decision-making process. For example, public consultation meetings or written submissions provide avenues for community members to express their views on potential project impacts. When these inputs are actively considered and addressed in the final project design, it creates a sense of ownership and collaboration. Take, for example, a proposed highway construction project that would displace several communities. Through the EIA process, alternative routes might be evaluated, and solutions such as constructing noise barriers or improving relocation packages could be implemented based on public feedback. This inclusive approach not only enhances the project's social acceptability but also reduces conflicts and opposition, leading to smoother project execution and higher overall satisfaction.

Maintenance of biodiversity and ecosystem stability is a crucial outcome of well-conducted EIAs. Biodiversity is vital for ecosystem resilience, providing services such as pollination, water purification, and climate regulation. EIAs help identify potential threats to biodiversity and propose measures to mitigate these risks. For instance, if a project is likely to disrupt wildlife corridors, an EIA might recommend creating green bridges or underpasses to ensure animal movement is not hindered. Furthermore, EIAs assess cumulative impacts, considering the collective effect of multiple projects on ecosystems over time. This holistic approach allows for better planning and management of environmental resources. In situations where large-scale developments are planned, such as dam construction or mining operations, EIAs play a pivotal role in ensuring that critical habitats are preserved and ecosystem functions remain intact. By maintaining biodiversity and ecosystem stability, EIAs contribute to the long-term health and sustainability of our natural environment.

Stakeholder Involvement

Public participation plays a pivotal role in the Environmental Impact Assessment (EIA) process. This subpoint delves into how inclusiveness, stakeholder engagement, transparency, and public opinion influence can significantly enhance EIA outcomes, making them more robust, credible, and effective.

Inclusiveness is fundamental to capturing diverse perspectives and concerns. Effective public participation means actively reaching out to representatives from various communities and demographics, including under-represented and marginalised groups. By incorporating a wide range of views, EIAs can identify potential environmental impacts that may not be immediately evident to project planners or experts. For instance, local residents might have first-hand knowledge of seasonal flooding that could affect a proposed construction site, information that might otherwise be overlooked. Inclusiveness ensures that the EIA process considers all relevant factors, leading to more comprehensive assessments and better-informed decisions.

By utilizing the knowledge and personal interests of those who are most directly affected by the projects, stakeholder engagement improves EIA outcomes. When stakeholders such as local businesses, environmental groups, and community leaders are involved, they bring valuable insights into the assessment process. This collaboration often results in more practical and sustainable project designs. Engaging stakeholders also helps in identifying potential socio-economic impacts early on, enabling mitigation strategies that address both environmental and community concerns. Effective stakeholder engagement fosters a sense of ownership and support for the project, which can be crucial for its successful implementation.

Transparency is essential in fostering trust between the sponsoring agencies and the public. Transparent processes involve sharing all relevant information, including data, criteria, and deliberations, with the public. Without transparency, public input may be based on incomplete or misunderstood information, leading to scepticism and opposition. Transparency ensures that all stakeholders are aware of how decisions are made and what factors are considered. This openness not only builds trust but also encourages more constructive engagement from the public. People are more likely to participate and support the EIA process if they feel confident that their contributions are valued and taken seriously. In fact, a transparent EIA process reduces the likelihood of conflict and increases the chances of project success by aligning stakeholder expectations with project realities (US EPA, 2014).

The influence of public opinion on project redesign via EIA processes cannot be understated. Public participation allows for feedback that can lead to significant modifications in project plans. For example, if a planned infrastructure project poses potential risks to a local water source, public opposition might prompt the project developers to consider alternative sites or advanced mitigation measures. This iterative process of receiving and integrating public feedback ensures that the final project design is not only technically sound but also socially acceptable. Moreover, public input can highlight innovative solutions that might not have been considered by the project team. This collaborative approach ultimately leads to more resilient and adaptive project outcomes.

A case study highlighting successful public participation in an EIA process can illustrate these points effectively. In the development of a new highway in a densely populated area, extensive public consultations were conducted at multiple stages. The project team organised community meetings, distributed surveys, and set up online platforms for feedback. This inclusive approach revealed concerns about noise pollution, displacement of residents, and potential harm to a nearby wetland. Based on this feedback, the project plan was adjusted to include noise barriers, compensation for displaced residents, and a modified route to protect the wetland. As a result, the project received broad community support and proceeded smoothly through the approval stages.

Another example can be drawn from a renewable energy project where stakeholder engagement played a critical role. During the planning phase of a wind farm, local farmers and environmental activists were invited to share their insights. Their input led to adjustments in turbine placement to minimise impacts on farmland and bird migration paths. Additionally, the project included community benefits such as shared revenue models and job creation programmes, further strengthening local support. This proactive engagement not only addressed environmental and social concerns but also built a strong partnership between the project developers and the community.

Transparency in EIA processes can be exemplified by a mining project where all assessment reports, decision-making criteria, and meeting minutes were made publicly available online. Regular updates and open forums were held to discuss progress and concerns. This level of openness ensured that all stakeholders had access to the same information, reducing misunderstandings and building trust. When a potentially significant environmental impact was identified, the project team worked collaboratively with concerned citizens and experts to develop effective mitigation strategies. This transparent approach not only enhanced the credibility of the EIA but also resulted in a more sustainable project design.

Case Studies Showcasing EIA Significance

Environmental Impact Assessments (EIAs) are critical tools used to evaluate the environmental consequences of proposed projects, ensuring that decisions made today do not compromise future generations. Real-world examples highlight the importance and efficacy of EIAs in various sectors, showcasing their role in promoting sustainable development.

Case studies provide powerful insights into the successful application of EIA processes. One notable instance is the Trans-Alaska Pipeline System (TAPS), where an extensive EIA before construction helped identify and mitigate potential environmental risks. The pipeline, spanning 800 miles, crosses sensitive ecosystems. Through rigorous assessment, measures were implemented to minimise impacts on wildlife, including caribou migration patterns. This proactive approach not only preserved biodiversity but also set a precedent for corporate responsibility in large-scale infrastructure projects.

In contrast, projects lacking thorough EIAs often face significant challenges. An example is the Belo Monte Dam in Brazil, which experienced considerable setbacks due to inadequate environmental assessments. Initially, the project overlooked key ecological concerns, resulting in severe deforestation, disruption of local communities, and loss of biodiversity. Protests and legal battles ensued, leading to delays and increased costs. This case underscores the necessity of comprehensive EIAs to foresee and address potential negative outcomes proactively.

Comparative analysis across different sectors reveals varying effectiveness of EIAs. In the energy sector, offshore wind farms have benefited immensely from detailed environmental assessments. For instance, the Hornsea One project in the UK, one of the world's largest offshore wind farms, underwent extensive EIA procedures to understand its impact on marine life and coastal ecosystems. The findings guided strategic siting and operational practices, mitigating adverse effects and enhancing project acceptance among stakeholders.

Conversely, the mining industry presents mixed results regarding the effectiveness of EIA. Initial assessments are inadequate for projects like the Ok Tedi Mine in Papua New Guinea, which is characterized by environmental degradation. Tailings disposal into rivers caused widespread pollution, affecting aquatic life and local communities. However, improved EIA practices in recent years, incorporating lessons learnt from such failures, have seen more sustainable mining operations emerge worldwide.

Innovative tools and approaches are continually evolving within EIA practices, enhancing their scope and accuracy. Geographic Information Systems (GIS) and remote sensing technologies now play pivotal roles in environmental assessments. These tools offer precise spatial data and real-time monitoring capabilities, ensuring that potential impacts are identified early and mitigated effectively. For instance, GIS was instrumental in planning the Cross rail project in London, facilitating the integration of environmental considerations in urban infrastructure development.

Moreover, strategic environmental assessment (SEA) represents a holistic approach, addressing the cumulative environmental effects of policies, plans, and programmes. SEAs complement project-specific EIAs by considering broader contexts and long-term sustainability goals. Sweden's SEA framework for national transport plans exemplifies this approach, aiming to reduce greenhouse gas emissions and promote green mobility.

Another innovative practice is the incorporation of ecosystem services valuation in EIAs. This methodology assesses the benefits provided by natural ecosystems, such as water purification, flood regulation, and recreational opportunities. By quantifying these services, decision-makers can better appreciate the trade-offs involved in development projects and prioritise conservation efforts. The restoration of the Florida Everglades, guided by ecosystem service valuation, demonstrates how integrating this concept into EIAs can lead to ecologically and economically sound outcomes.

Public participation forms a crucial component of effective EIAs, fostering transparency and inclusiveness. Engaging local communities and stakeholders ensures that diverse perspectives and concerns are addressed, enhancing the credibility and acceptance of the assessment process. The Djerdap Hydropower Plant project in Serbia illustrates successful stakeholder engagement, where open dialogues with affected populations led to modifications in the project's design, aligning it with community needs and environmental standards.

While traditional EIAs focus on preventing harmful impacts, adaptive management approaches introduce flexibility to respond to unforeseen changes during project implementation. Adaptive management involves continuous monitoring, evaluation, and adjustment of mitigation measures based on real-world outcomes. The Snowy Mountains Hydro-Electric Scheme in Australia employs adaptive management to balance water resource utilisation with environmental preservation, demonstrating its potential in dynamic and complex projects.

Emerging trends also emphasise the integration of climate change considerations into EIA processes. Climate risk assessments identify vulnerabilities and guide adaptation strategies, ensuring that projects remain resilient in the face of changing climate conditions. The Thames Estuary 2100 Plan in the UK incorporates climate scenarios in its EIA, addressing future flood risks and safeguarding urban infrastructure against rising sea levels.

Incorporating social impact assessments (SIAs) alongside EIAs ensures a comprehensive understanding of the project's implications on both the environment and society. SIAs consider factors such as community health, livelihoods, and cultural heritage, providing a balanced view of development impacts. The Merowe Dam project in Sudan faced significant criticism for neglecting social dimensions, resulting in displacement and socio-economic disruptions. Integrating SIAs could have facilitated more equitable and acceptable outcomes.

In summary, real-world examples underscore the importance and impact of effective EIAs in diverse sectors. Case studies reveal successes in mitigating environmental risks and enhancing project sustainability while highlighting the pitfalls of inadequate assessments. Comparative analyses across sectors demonstrate varied EIA effectiveness, identifying areas for improvement. Innovative tools and approaches, such as GIS, SEA, ecosystem services valuation, and adaptive management, enhance EIA robustness. Public participation and the integration of climate change and social considerations further strengthen the EIA process, ensuring that development projects contribute positively to sustainable development goals.

Summary and Reflections

This chapter has laid a solid foundation for understanding Environmental Impact Assessments (EIA) by delving into their fundamental concepts, purposes, and processes. It has emphasised the importance of EIAs in identifying potential environmental impacts early in project planning, allowing for informed decision-making and the mitigation of adverse effects. The collaborative nature of EIAs, involving various experts and stakeholders, ensures a robust and comprehensive evaluation that addresses multiple dimensions of the environment.

By incorporating environmental considerations into project design and execution, EIAs play a pivotal role in promoting sustainable development. They aid stakeholders in balancing economic, social, and environmental factors, ultimately leading to more responsible and accepted project outcomes. The chapter has also highlighted the significance of public participation and transparency in enhancing the credibility and effectiveness of EIAs. Moving forward, it is crucial for all involved stakeholders to continue leveraging these assessments to ensure the long-term sustainability and health of our natural environments.

Reference List

Abdulhakim Aljareo, Watson, I., & Schwaibold, U. (2023, April 6). Developing an evaluation approach to consider the influence of country context on Environmental Impact Assessment performance from a southern African perspective. https://doi.org/10.1002/ieam.4771

Environmental Impact Assessment Advantages And Disadvantages. (2023, November 29). Cypress Environment and Infrastructure. https://cypressei.com/engineering/environmental-impact-assessment-advantages-disadvantages/

Environmental Impact Assessment: A Comprehensive Guide - ESG Research Pro. (2023, November 22). Esgresearch.pro. https://esgresearch.pro/environmental-impact-assessment/

Envchemp. (2023, July 12). 9: Importance of Environmental Impact Assessment. Envirochem Patnas Limited. https://envchempatnas.com/9-importance-of-environmental-impact-assessment/

ESS Topic 1.4: Sustainability. (n.d.). AMAZING WORLD of SCIENCE with MR. GREEN. https://www.mrgscience.com/ess-topic-14-sustainability.html

González, A., Therivel, R., Lara, A., & Lennon, M. (2023, July). Empowering the public in environmental assessment: Advances or enduring challenges? Environmental Impact Assessment Review. https://doi.org/10.1016/j.eiar.2023.107142

Nita, A., Fineran, S., & Rozylowicz, L. (2022, January 1). Researchers' perspective on the main strengths and weaknesses of Environmental Impact Assessment (EIA) procedures. Environmental Impact Assessment Review. https://doi.org/10.1016/j.eiar.2021.106690

RJ. (2023, September 28). Environmental Impact Assessments: Your Guide to Responsible Development. Equator. https://equatorstudios.com/gis-data-and-analysis/environmental-impact-assessments/

US EPA. (2014, February 24). Public Participation Guide: Introduction to Public Participation US EPA. US EPA. https://www.epa.gov/international-cooperation/public-participation-guide-introduction-public-participation

Zhao, Y., Li, X., Mo, H., Zhan, L., Yao, Y., Li, Y., & Li, H. (2023, July 1). How does the environmental impact assessment (EIA) process affect environmental performance? Unveiling EIA effectiveness in China: A practical application within the thermal power industry. Environmental Impact Assessment Review. https://doi.org/10.1016/j.eiar.2023.107120

Chapter 2

Key Elements of the EIA Process

Understanding the key elements of the Environmental Impact Assessment (EIA) process is essential for conducting thorough evaluations of potential environmental effects. This chapter delves into the critical stages and components that make up an effective EIA, ensuring that professionals, policymakers, and students gain comprehensive insights into this structured approach to environmental protection. By examining these key elements, one can appreciate the intricacies involved in predicting, assessing, and mitigating adverse environmental impacts stemming from various projects.

In this chapter, readers will explore the sequential stages of the EIA process, starting with screening to determine whether a full assessment is necessary. The discussion then progresses to detailed steps such as scoping, which defines the parameters and key issues for evaluation. The collection of baseline data is emphasized, highlighting its importance in establishing existing conditions before project development begins. Following this, the focus shifts to impact analysis, where potential environmental changes are identified and assessed. Additionally, the chapter addresses common challenges faced during the EIA process and showcases the importance of standardized guidelines, transparency, public participation, and stakeholder engagement.

Screening

Screening is the initial step in determining whether a proposed project requires an Environmental Impact Assessment (EIA). This process is essential for identifying projects that might have significant effects on the environment and ensuring that appropriate levels of assessment are conducted. The main goal of screening is to decide if a full EIA is necessary, thus saving time and resources by focusing efforts on projects with the potential for substantial environmental impacts.

Several factors influence the outcomes of the screening process. One of the primary considerations is the size and scale of the project. Large-scale projects that involve extensive construction or land alteration are more likely to require a full EIA due to their potential for considerable environmental disruption. For example, constructing a large dam may necessitate a comprehensive EIA because of the likely significant impacts on local ecosystems, water flow, and communities.

Location is another critical factor affecting screening outcomes. Projects situated in environmentally sensitive areas, such as wetlands, forests, or regions with endangered species, often require more rigorous assessment. Environmental sensitivity can significantly increase the likelihood that a full EIA will be mandated. For instance, developing infrastructure in a coastal zone may be subject to stringent screening processes due to the high ecological value and vulnerability of marine ecosystems.

The sensitivity of the surrounding environment also plays a vital role in the screening process. Areas that are already under environmental stress or degradation might need detailed EIAs for new projects, even if they are relatively small in scale. This is to ensure that additional pressures do not exacerbate existing problems. An example could be industrial development in a region experiencing severe air pollution. Even minor projects in such areas could necessitate thorough scrutiny to prevent further degradation.

Common criteria used to determine the necessity of an EIA include evaluating the potential for adverse environmental impacts and gauging public interest and concern. Assessing potential environmental impacts involves predicting how a project might affect various aspects of the environment, such as air and water quality, biodiversity, and human health. Public interest is also a crucial criterion since projects that generate significant public concern or opposition are likely to undergo more comprehensive assessments. An example of this can be seen in urban development projects which often face intense public scrutiny, leading authorities to favor conducting full EIAs to address community concerns comprehensively.

The results of the screening process can lead to different outcomes, each with distinct implications for the project's development. One possible result is that a full EIA is required, indicating that the project has the potential for significant environmental impacts and needs detailed study and mitigation planning. This outcome typically means that the project will undergo a more prolonged review process, involving various technical studies and public consultations. The objective is to identify, predict, and mitigate adverse environmental effects before project approval.

Alternatively, the screening process might conclude that only a minimal assessment is needed. This is often the case for smaller projects or those situated in less sensitive environments where the potential impacts are considered negligible or easily manageable. Projects requiring minimal assessment typically follow a less rigorous review pathway, allowing for quicker decision-making while still ensuring that environmental considerations are addressed.

To illustrate, a small-scale community-based solar power installation may not necessitate a full EIA if it is located in a non-sensitive area and poses minimal environmental risks. Instead, a simplified assessment focusing on specific aspects like noise or visual impact might suffice. Conversely, a large mining operation in a remote, biodiverse region would almost certainly trigger a full EIA due to its substantial environmental footprint and the complexity of mitigating its impacts.

Understanding these varied outcomes is crucial for environmental professionals, policymakers, and stakeholders involved in project development. It ensures that appropriate measures are taken to balance developmental needs with environmental protection. By recognizing the importance of the screening process and the factors influencing its outcomes, decision-makers can make informed choices about the level of assessment required for different projects.

Scoping

Understanding the scoping phase in an Environmental Impact Assessment (EIA) is essential for a thorough evaluation of a project's potential environmental and socio-economic impacts. Scoping helps to establish clear boundaries, identify key issues, and engage relevant stakeholders, thereby setting the stage for a comprehensive EIA.

At its core, scoping focuses on defining the parameters of the EIA. This step ensures that only pertinent issues are investigated, thus optimizing resources and time. Determining these boundaries involves assessing the extent of the geographical area, the temporal scope, and the specific environmental aspects to be considered. For instance, a coastal development project might include a focus on marine ecosystems, shoreline stability, and local fishing communities as part of its scope. By narrowing down the key issues, scoping helps avoid diluting efforts on less significant concerns.

The involvement of stakeholders during the scoping phase significantly enriches the process. Stakeholders include individuals, groups, and organizations affected by or interested in the project. Their engagement brings diverse perspectives and localized knowledge, reflecting the concerns and priorities of those directly impacted. Methods such as public meetings, surveys, and workshops facilitate this involvement. For example, in a proposed industrial project near a residential area, engaging with local residents can uncover issues like air quality, noise pollution, and traffic congestion that may not be immediately apparent to project developers. Incorporating stakeholders' inputs not only enhances the relevance of the EIA but also fosters transparency and trust.

Identifying critical environmental and socio-economic factors is another vital aspect of scoping. These factors form the basis for prioritizing which issues merit detailed analysis. Environmental factors could include biodiversity, water quality, and land use, while socio-economic factors might encompass community health, economic livelihoods, and cultural heritage. Using criteria such as magnitude, duration, and reversibility of impacts helps prioritize these factors. For instance, a large-scale mining project might prioritize analyzing impacts on water resources due to their significance for both the ecosystem and local communities. Prioritization ensures that the most pressing issues receive the attention they deserve, leading to more effective mitigation strategies.

An overview of the scoping report provides insight into its significance and typical content. The scoping report is a document that outlines the scope of the EIA, detailing the identified key issues, stakeholder inputs, and the rationale for prioritizing certain factors. This report serves as a roadmap for conducting the EIA, guiding subsequent data collection, analysis, and impact assessment phases. Typical content includes an introduction, objectives of the scoping exercise, a description of the project, stakeholder engagement results, and prioritized environmental and socio-economic factors. Additionally, it might contain recommendations for further studies or baseline data requirements. For instance, if air quality emerged as a critical issue during scoping, the report might recommend specific air quality monitoring protocols to ensure comprehensive data collection.

To ensure clarity, the scoping report should be well-structured and written in plain language. Clear headings, concise summaries, and visual aids like maps and charts enhance readability and comprehension. This is particularly important for non-technical stakeholders who need to understand the scope of the EIA and how their concerns have been addressed. Recommendations for ensuring clarity might include using bullet points for key information, providing a glossary of technical terms, and including an executive summary for quick reference. A well-crafted scoping report not only facilitates the EIA process but also serves as a communication tool, demonstrating the thoroughness and transparency of the scoping phase.

Baseline Data Collection

The collection and analysis of baseline data is a critical first step in the Environmental Impact Assessment (EIA) process. This comprehensive approach ensures that environmental professionals, policymakers, and scholars have a clear understanding of the existing conditions within the project area before any development begins. Baseline data serves multiple functions: it provides a reference point for future changes, establishes ecological health indicators, and helps in predicting potential impacts of proposed activities.

Baseline data collection typically starts with identifying relevant ecological, socio-economic, and cultural parameters indicative of the area's current state. These parameters might include water quality, air quality, soil health, biodiversity indices, population demographics, and land-use patterns. Having precise and comprehensive baseline data is vital because it forms the foundation upon which the entire EIA process is built. Without this information, accurately assessing the significance of potential environmental impacts would be challenging.

Different methodologies are employed to collect baseline data, each with its own set of advantages and limitations. Field studies, which involve direct observation and sample collection, are one of the most common methods. These studies often provide highly specific and detailed data, such as species counts, pollutant levels, and habitat conditions. For example, biologists might conduct surveys to inventory local wildlife, or chemists may analyze water samples to determine pollutant concentrations.

Remote sensing technology offers another valuable tool for baseline data collection. Using satellite imagery, aerial photography, and other remote-sensing techniques, researchers can gather large-scale environmental data over extended periods. This method is particularly effective for monitoring land cover changes, assessing vegetation health, and detecting geographic features that are not easily accessible for ground-based surveys. For instance, changes in forest cover due to deforestation or urban expansion can be monitored via satellite images, providing a continuous record of land-use changes over time.

Existing databases can also serve as a rich source of baseline data. Many governmental and non-governmental organizations maintain extensive records on various environmental parameters. Utilizing these pre-existing datasets can save time and resources while still providing a robust foundation for the EIA. For example, national meteorological agencies often collect and publish long-term climate data that can be invaluable for understanding historical weather patterns and predicting future climatic changes.

Engaging local communities in the data collection process can enhance the specificity and relevance of the baseline data. Local knowledge often includes insights into seasonal variations, rare species, culturally significant sites, and other nuanced information that might not be readily apparent through conventional scientific methods alone. Incorporating such knowledge fosters a trust-building relationship between project developers and the community, leading to more accurate and context-sensitive data. Additionally, participatory approaches ensure that the community's concerns and values are reflected in the assessment outcomes.

Analyzing the collected baseline data must be done meticulously to maintain transparency and rigor. Clear documentation of methodologies, sampling protocols, and analytical techniques is essential to ensure that the findings are credible and reproducible. One key aspect of rigorous analysis is the use of appropriate statistical tools to identify trends, correlations, and anomalies within the data. This statistical rigor helps in quantifying environmental parameters and inferring their significance relative to the project's potential impacts.

Transparency in data analysis involves making all data sources, methodologies, and interpretations publicly available whenever possible. This openness allows external experts to validate the findings, contributing to the credibility of the EIA process. For example, publishing detailed reports on water quality assessments, including raw data and analytical procedures, enables other scientists to replicate the study and verify its conclusions.

Addressing potential biases during data collection and analysis is another crucial component. Biases can arise from various sources, including sampling locations, timing, and observer experience. Implementing standardized protocols and training for field personnel can help minimize these biases. Moreover, peer reviews and independent audits of data collection and analysis procedures provide an additional layer of quality assurance.

For instance, consistent sampling methods across different times and locations ensure that data comparisons are valid and meaningful. If studying aquatic ecosystems, taking water samples at the same depth and location across different seasons removes variability that could skew results. Furthermore, regularly calibrating equipment and cross-verifying data with other sources can highlight and correct discrepancies.

Impact Analysis

Impact analysis forms the core of Environmental Impact Assessment (EIA) findings, providing the necessary framework to ensure the validity and reliability of the assessment. Once baseline conditions are established, impact analysis can begin. This process is vital as it identifies potential changes a project might induce in the environment. The analysis serves as a foundation for decision-making, allowing stakeholders to understand how proposed activities may alter ecological, social, and economic structures.

The types of impacts analyzed during an EIA can be broadly categorized into direct, indirect, and cumulative impacts. Direct impacts result from immediate interactions between the project and the environment, such as habitat destruction due to construction. Indirect impacts are secondary effects that occur over time, often resulting from shifts in environmental or social parameters prompted by the primary activities. For example, pollution runoff from construction sites can eventually affect water quality in nearby rivers. Cumulative impacts consider the aggregate effect of multiple projects or actions over time, which could lead to substantial environmental changes even if individual actions appear insignificant. Analyzing these impacts ensures a holistic understanding of a project's overall footprint.

Ecological impacts refer to changes in biodiversity, habitat integrity, and ecosystem services. Social impacts include demographic shifts, cultural disruptions, and changes in land use patterns.

Economic impacts cover aspects like employment opportunities, impacts on local businesses, and resource allocation. Each of these categories must be carefully examined to present a comprehensive picture of a project's impact. For instance, building a factory may provide jobs but could also disrupt local wildlife habitats and displace communities.

Several tools and methodologies are employed in predicting these impacts, ranging from qualitative to quantitative techniques. Qualitative methods often involve expert judgment, stakeholder consultations, and scenario analysis to predict possible outcomes. These methods are beneficial when data is scarce or when evaluating non-quantifiable factors like community well-being. Quantitative techniques, on the other hand, utilize statistical models, Geographic Information Systems (GIS), and computer simulations to provide numerical estimates of potential impacts. These techniques are advantageous for assessing measurable factors such as emissions levels, deforestation rates, or changes in water quality.

Modeling tools are particularly useful for visualizing and predicting future scenarios. Software like Stella or Vensim allows analysts to create dynamic models that simulate environmental systems. These models can incorporate various data points, enabling the simulation of different scenarios to understand potential impacts better. For example, hydrological models can predict how construction near a river might influence water flow and flood risks.

Once impact predictions are made, linking them to mitigation planning becomes crucial. Mitigation measures aim to avoid, minimize, or compensate for adverse effects. Effective mitigation planning often stems from thorough impact analysis, as understanding potential impacts enables the development of targeted strategies to address them. One example of successful mitigation derived from robust impact analysis is the installation of wildlife corridors in infrastructure projects. These corridors enable animals to safely traverse fragmented habitats, reducing the risk of roadkill and promoting biodiversity.

Another example is the implementation of green buffer zones around industrial areas to filter out pollutants before they reach residential zones or natural habitats. Such measures not only protect human health and the environment but also enhance the sustainability and acceptance of the project among local communities. Adaptive management practices also play a significant role in mitigation. By continuously monitoring impacts and adjusting strategies as needed, adaptive management helps ensure that the mitigation measures remain effective over time.

Process Challenges

Inconsistent guidelines can significantly impact the effectiveness of Environmental Impact Assessments (EIAs). Without clear and consistent regulatory frameworks, project assessments may vary greatly, leading to unreliable data and conclusions. Inconsistencies can arise from differences in local, regional, or national regulations, causing confusion among stakeholders and reviewers. To mitigate this issue, it is essential to establish standardized guidelines that ensure uniformity and reliability across all EIA processes. This includes defining specific criteria for assessing environmental impacts, setting precise methodologies for data collection, and detailing requirements for public participation. Such measures not only enhance the credibility of the EIA process but also provide a clear roadmap for practitioners, ensuring that all assessments are conducted under uniform principles.

Transparency and public participation play vital roles in ensuring credible EIA processes. When project details are openly shared with the public, it fosters trust and accountability. Public involvement allows for a broader range of perspectives, identifying potential impacts that experts might overlook. Effective public participation requires accessible and clear communication channels, timely dissemination of information, and opportunities for feedback at multiple stages of the EIA process. By promoting transparency, EIA practitioners build community trust, reducing opposition and facilitating smoother project implementation. Moreover, public input can lead to more comprehensive assessments, as local communities often possess valuable insights into their environment that can enrich the analysis.

Stakeholder input is crucial throughout the EIA process, shaping decisions and contributing to more robust outcomes. Stakeholders, including local communities, government agencies, NGOs, and industry representatives, bring diverse viewpoints and expertise to the table. Engaging these groups early and consistently helps identify issues of concern, mitigates risks, and enhances the overall quality of the assessment. One effective approach is to create stakeholder engagement plans that outline methods for involving various groups, such as workshops, surveys, public hearings, and focus groups. These interactions help gather a wide array of opinions and data, ensuring that the EIA addresses the interests and concerns of all affected parties. Furthermore, involving stakeholders in decision-making processes ensures that the EIA findings are well-rounded and reflective of collective insights.

Case studies highlighting poor choices and their repercussions provide valuable lessons for EIA practitioners. For instance, consider a scenario where inadequate baseline data collection led to an underestimation of a project's impact on a critical habitat. The resulting damage sparked public outrage and legal challenges, delaying the project and increasing costs. Such cases stress the importance of thorough and accurate data collection during the initial phases of the EIA. Another example might involve a lack of stakeholder engagement, which resulted in significant opposition from local communities and subsequent project delays. These situations illustrate how overlooking key elements of the EIA process can have severe consequences, emphasizing the need for careful consideration and adherence to established guidelines at every phase.

To overcome common challenges in EIAs, it is essential to address inconsistent guidelines by advocating for clear regulatory frameworks.

Standardized protocols should be developed and implemented, ensuring consistency across different jurisdictions. Training programs for EIA practitioners can also help maintain high standards by familiarizing them with best practices and updated regulations. Additionally, fostering a culture of continuous improvement, where feedback from past projects is used to refine guidelines and procedures, can lead to more effective and reliable assessments.

Promoting transparency and public participation requires deliberate efforts to engage with communities and stakeholders throughout the EIA process. Tools such as digital platforms, social media, and public information sessions can facilitate these engagements, making it easier for individuals to access information and voice their opinions. Ensuring that public comments and suggestions are genuinely considered in the decision-making process demonstrates respect for community input and reinforces the legitimacy of the EIA.

Moreover, creating formal mechanisms for stakeholder consultation can systematically integrate diverse viewpoints into the EIA process. For example, establishing advisory committees comprising representatives from various stakeholder groups can provide a structured platform for ongoing dialogue. This approach not only enriches the EIA with diverse knowledge but also helps build consensus and address conflicts early on.

Learning from case studies involves analyzing past projects to understand the pitfalls and successes of previous EIAs. Detailed reviews of these case studies can highlight specific areas for improvement, offering practical insights that inform future assessments. By documenting and sharing these lessons, the EIA community can collectively enhance its practices and avoid repeating past mistakes.

Summary and Reflections

In this chapter, we explored the various stages and components critical to conducting an effective Environmental Impact Assessment (EIA). We began with screening, which helps determine whether a project requires a full EIA, focusing on factors like project size, location, and environmental sensitivity. Next, we delved into scoping, where key issues are identified, stakeholders are engaged, and the boundaries of the assessment are defined. This step ensures that only relevant concerns are investigated, optimizing both resources and time.

We also examined the importance of collecting baseline data, which provides a reference point for evaluating future changes and potential impacts. Various methodologies, from field studies to remote sensing, help gather this essential information. Finally, we discussed impact analysis, examining direct, indirect, and cumulative effects of the proposed projects, and linking these findings to mitigation strategies. Understanding these processes enables environmental professionals, policymakers, and students to conduct more robust assessments, balancing developmental needs with environmental protection.

Chapter 3

Technological Tools in EIA

Technological tools have brought a transformative edge to Environmental Impact Assessments (EIAs). By incorporating modern technologies, environmental professionals can enhance the accuracy and effectiveness of their assessments. This chapter delves into the various technological advancements that have been integrated into EIAs, highlighting how these tools are revolutionising the assessment process.

The chapter provides a comprehensive overview of Geographic Information Systems (GIS), remote sensing techniques, environmental modelling, data analytics, and stakeholder engagement tools. These technologies facilitate detailed spatial analysis, real-time data monitoring, predictive modelling, and inclusive stakeholder participation. Through concrete examples and practical applications, readers will gain insights into how these tools improve environmental assessments and support sustainable development.

Geographic Information Systems (GIS)

Geographic Information Systems (GIS) have become indispensable in environmental impact assessments (EIAs) due to their ability to facilitate spatial analysis and decision-making. By leveraging GIS technology, environmental professionals can visualise complex data, conduct detailed spatial analyses, model future scenarios, and engage stakeholders effectively.

One of the primary advantages of GIS is its robust data visualisation capabilities. By transforming intricate environmental data into visual maps and charts, GIS ensures that stakeholders can easily interpret and comprehend the findings. For example, a map showing pollution levels across different regions can quickly highlight hotspot areas needing urgent attention. This visual representation makes it easier for policymakers to grasp the severity and spread of environmental issues, aiding them in formulating targeted interventions. According to England (2024), GIS links alphanumeric data to cartographic layers, making it a reliable tool for analysing and visualising territorial impacts. This capacity for visualisation fosters better communication between experts and non-experts, promoting more informed decision-making.

Beyond visualisation, GIS offers powerful spatial analysis tools essential for assessing environmental impacts. These tools enable analysts to overlay various datasets—such as land use, hydrology, and population density—to identify correlations and potential conflicts. For instance, by layering soil quality data with agricultural land use patterns, GIS can help determine which farming practices are sustainable and which may cause soil degradation. Additionally, GIS supports statistical analyses like proximity analysis and buffer zone creation. Proximity analysis can reveal how close a planned development is to sensitive habitats or protected areas, while buffer zones can show areas that need safeguarding from certain activities. These analytical capabilities allow for precise evaluations of environmental impacts, leading to more effective mitigation strategies.

Scenario planning is another area where GIS significantly enhances EIA processes. By integrating various data layers, GIS can simulate different development scenarios and predict their potential environmental outcomes. For example, GIS can model how urban expansion might affect local water bodies under varying conditions, such as increased rainfall or drought. These simulations help planners and developers understand the long-term implications of their projects and choose the most sustainable options. Scenario planning using GIS also enables the assessment of cumulative impacts, providing a holistic view of how multiple projects might interact and affect the environment over time. This forward-looking approach ensures that decisions made today do not lead to unintended adverse consequences in the future.

Moreover, GIS excels in facilitating stakeholder interaction, an essential component of transparent and inclusive decision-making. Engaging the community through GIS tools helps build public trust and ensures that all voices are heard in the assessment process. Interactive GIS maps and dashboards allow community members to see real-time data and provide input, making them active participants in environmental governance. For example, a public GIS platform can show the projected impacts of a new highway on local wildlife habitats, enabling residents to voice their concerns or support based on solid data. This transparency demystifies the EIA process and fosters a sense of ownership among stakeholders, leading to more robust environmental outcomes.

The use of GIS in EIAs exemplifies how technology can bridge the gap between complex data and practical decision-making. By presenting environmental information visually, supporting detailed spatial analyses, enabling scenario planning, and fostering stakeholder engagement, GIS transforms the way environmental assessments are conducted. It helps professionals and policymakers make well-informed decisions that balance development needs with environmental preservation.

Remote Sensing Techniques

Remote sensing methods have become essential in Environmental Impact Assessments (EIAs), providing extensive data for monitoring and evaluating environmental conditions. These techniques offer various applications, from satellite imagery to aerial surveys, change detection, and integration with Geographic Information Systems (GIS). Together, they enhance the scope and accuracy of EIAs, ensuring better-informed decision-making processes.

Satellite imagery plays a pivotal role in remote sensing by providing comprehensive views of the Earth's surface. This technology leverages satellites equipped with advanced sensors to capture data over vast areas consistently. Satellite imagery is particularly valuable in detecting land use changes and their subsequent environmental impacts. By analysing spectral signatures from different time periods, it is possible to identify variations in vegetation cover, urban expansion, deforestation, and other significant alterations in land use patterns. For instance, policymakers can rely on satellite data to gauge the extent of deforestation and implement necessary conservation measures. This method also allows for continuous monitoring, making it easier to observe trends and immediate shifts in environmental conditions.

Aerial surveys complement satellite imagery by offering high-resolution data at specific sites. Drones and manned aircraft equipped with advanced cameras and sensors are employed to capture detailed images and videos. The advantage of using drones lies in their ability to hover over target areas, providing close-up views that satellite imagery cannot achieve due to its broader spatial coverage. Aerial surveys are especially useful in assessing areas prone to environmental degradation, such as mining sites or regions affected by natural disasters like floods and landslides. High-resolution aerial photos enable precise mapping of affected zones, leading to more effective mitigation strategies. Additionally, drones can access hard-to-reach locations, offering unique perspectives that enhance understanding and responses to environmental challenges.

Change detection is another vital aspect of remote sensing that involves analysing and quantifying changes in the environment over time. This technique utilises multi-temporal remote sensing data to detect differences between images taken at different times. Change detection algorithms can highlight alterations in landscape features, such as shifts in vegetation health, water bodies, and soil erosion rates. By identifying these changes, stakeholders can assess the effectiveness of implemented environmental management practices and make informed adjustments. For example, a Before-After Control Impact (BACI) analysis using satellite imagery has demonstrated significant improvements in vegetation conditions of projects aimed at soil and water conservation (Vorovencii,

2011). Such analyses help validate project outcomes and guide future initiatives.

Integration with GIS enhances the utility of remote sensing data by enabling comprehensive analysis and visualisation. GIS platforms allow users to merge remote sensing data with other spatial datasets, creating layered maps that provide a holistic view of environmental conditions. This integration facilitates spatial analysis, where users can measure distances, areas, and spatial relationships between various environmental factors. For instance, combining satellite imagery with GIS data on land use planning and population density helps identify urban sprawl patterns and predict potential growth areas. GIS also supports modelling scenarios that can inform policy decisions. For example, officials can simulate the impact of proposed infrastructure developments on local ecosystems and devise better planning strategies to minimise adverse effects.

Furthermore, integrating remote sensing with GIS supports scenario development, which is crucial for anticipating the environmental consequences of proposed projects. Spatial modelling aids in predicting potential impacts, allowing stakeholders to explore alternative solutions and mitigation strategies. For example, a model might predict how a new highway could disrupt wildlife migration routes and propose alternative pathways to preserve habitats. These predictive capabilities are invaluable for proactive environmental management and policy formulation.

The integration of remote sensing and GIS also enhances stakeholder engagement in environmental assessments. Visual representations of data, such as thematic maps and 3D models, make complex information accessible to non-expert audiences. Stakeholders, including community members, policymakers, and developers, can participate meaningfully in discussions about environmental impacts and proposed interventions. Interactive GIS platforms enable stakeholders to view and comment on EIA reports, fostering transparency and inclusivity in the decision-making process. Public participation GIS (PPGIS) tools further encourage community involvement by allowing localised data sharing and feedback collection.

Environmental Modelling

Environmental Impact Assessments (EIAs) are crucial in ensuring that development projects proceed with minimal adverse effects on the environment. Environmental modelling serves as a fundamental tool in this process, helping practitioners understand and predict potential impacts. In this section, we will explore how predictive modelling, integrated modelling frameworks, scenario analysis, and uncertainty analysis support EIAs.

Predictive modelling is an essential component of environmental modelling. It involves the use of mathematical and computational techniques to forecast future environmental conditions under various scenarios. Predictive models can simulate how changes in variables like temperature, precipitation, or pollutant levels might affect ecosystems over time. For instance, a predictive model could estimate the impact of increased industrial activity on air quality in a metropolitan area, providing valuable data for planners and policymakers. By forecasting these conditions, predictive modelling allows EIA professionals to anticipate and mitigate negative environmental impacts before they occur.

Transitioning from predictive modelling, integrated modelling frameworks combine multiple factors into a single cohesive model. These frameworks take into account ecological, social, and economic dimensions, offering a holistic perspective of potential impacts. An example of such a framework is the Integrated Assessment Model (IAM), which assesses the interplay between human activities and environmental systems. By incorporating data from various domains, IAMs provide a comprehensive view of how a proposed project might affect not only the natural environment but also local communities and economies. This broader understanding supports more informed decision-making and helps balance development goals with sustainability objectives.

Scenario analysis plays a vital role in environmental modelling by allowing stakeholders to test different management options and evaluate potential outcomes. Through scenario analysis, various "what if" situations can be examined. For example, urban planners might use scenario analysis to explore the effects of different land-use strategies on regional water resources. By modelling different development pathways, it becomes possible to identify the most sustainable options. This method is particularly useful for assessing long-term impacts and developing adaptive management strategies that can respond to changing conditions.

One practical application of scenario analysis is in coastal management. Coastal areas are susceptible to a range of environmental threats, including sea-level rise, storm surges, and erosion. By using scenario analysis, decision-makers can evaluate different protective measures, such as constructing seawalls, restoring wetlands, or implementing zoning regulations. Each scenario is modelled to predict its effectiveness and potential side effects. This approach helps prioritise actions that offer the greatest benefit while minimising harm to the environment and communities.

Uncertainty analysis is another critical aspect of environmental modelling. Given the complexity of natural systems and the limitations of available data, uncertainties are inherent in any modelling process. Uncertainty analysis aims to quantify and address these uncertainties, providing a clearer picture of the risks involved. For instance, when predicting the impact of a new agricultural practice on soil health, uncertainties might arise from variations in soil composition, weather patterns, and crop responses. By identifying and analysing these uncertainties, EIA professionals can better understand the range of possible outcomes and develop strategies to manage risks effectively.

Incorporating uncertainty analysis into EIAs enhances the robustness and reliability of predictions. One technique used in uncertainty analysis is Monte Carlo simulation, which involves running multiple simulations with varying input parameters to generate a range of possible outcomes. This statistical approach helps to highlight the probability of different scenarios occurring, thus aiding decision-makers in accounting for uncertainty in their planning processes. Moreover, sensitivity analysis, which examines how changes in specific variables affect model outcomes, can pinpoint which factors have the most significant impact on results, guiding where further research or data collection may be needed.

The integration of these modelling techniques supports a proactive approach to environmental management. By leveraging predictive models, integrated frameworks, scenario analysis, and uncertainty assessments, EIA practitioners can make more informed decisions that consider both immediate and long-term impacts. This comprehensive approach not only helps in mitigating adverse environmental effects but also promotes sustainable development practices that benefit society as a whole.

Furthermore, the continuous advancement of technology is enhancing the capabilities of environmental modelling. Improved data collection methods, such as remote sensing and geographic information systems (GIS), provide more accurate and detailed environmental data. Advanced computing power enables the processing of large datasets and complex models in a shorter time frame, making it feasible to conduct more sophisticated analyses. These technological advancements ensure that environmental models are increasingly precise and relevant, thereby improving the quality of EIAs.

Data Analytics in Environmental Assessments

The rapid advancements in technology have significantly transformed many fields, including Environmental Impact Assessments (EIAs). One of the most influential developments is the role of data analytics in enriching EIAs by offering insightful analysis and decision support. This section delves into how data analytics can revolutionise environmental assessments through big data utilisation, real-time data analysis, machine learning applications, and risk assessment enhancement.

Big Data Utilisation:

Big data refers to the vast amounts of data generated from numerous sources, such as remote sensors, satellite imagery, social media, and more. By integrating these large datasets, environmental professionals can conduct more comprehensive and accurate assessments. For instance, big data can include diverse information such as biodiversity, weather patterns, pollution levels, and human activities, offering a holistic view of environmental conditions. The use of big data enables the identification of subtle yet significant environmental changes that might be missed through traditional data collection methods. Moreover, the integration of these datasets facilitates the detection of patterns and correlations, enhancing the predictive power of EIAs. This approach not only improves current assessments but also helps anticipate future environmental impacts.

Real-Time Data Analysis:

One of the notable advantages of modern analytical tools is their ability to perform real-time data analysis. Real-time analytics provide immediate insights into ongoing environmental conditions, enabling swift responses and timely interventions. For example, sensors placed in various ecosystems can continuously monitor parameters like air and water quality, providing instant feedback to stakeholders. Utilising real-time data ensures that any negative trends or sudden ecological disturbances are quickly identified and addressed, thereby mitigating potential adverse effects on the environment. Furthermore, this capability supports dynamic EIA processes where corrective measures can be implemented promptly, ensuring the sustainability of projects and minimising environmental damage.

Machine Learning Applications:

The application of machine learning algorithms in EIAs has opened new frontiers in environmental data analysis. Machine learning techniques can process vast datasets to identify trends, correlations, and outcomes that may be imperceptible through conventional analysis. For example, machine learning models can predict species' migration patterns based on historical data, climate conditions, and human activities, aiding in biodiversity conservation efforts. These algorithms can also help in monitoring pollution sources by analysing industrial emissions data and correlating them with health outcomes in nearby communities. As machine learning continues to evolve, its capacity to provide refined and precise assessments will improve, making EIAs more robust and reliable. This technology not only enhances the depth of analysis but also offers a proactive approach to environmental management by predicting potential impacts before they become critical issues.

Risk Assessment Enhancement:

Data analytics plays a crucial role in enhancing risk assessment within EIAs by identifying potential environmental risks and informing mitigation strategies. Advanced analytics can evaluate the likelihood and severity of environmental hazards, such as flooding, droughts, or chemical spills, resulting from proposed projects. By leveraging historical data, simulation models, and scenario analyses, it is possible to develop comprehensive risk profiles for different environmental factors. These profiles enable policymakers and project developers to devise effective mitigation plans that minimise risks while maximising project benefits. Additionally, data-driven risk assessments foster transparency and accountability, as stakeholders can see the evidence and rationale behind decisions, thus promoting public trust and acceptance of projects.

Stakeholder Engagement Tools

Technological tools play a fundamental role in facilitating stakeholder engagement during the Environmental Impact Assessment (EIA) process. These tools enhance transparency, inclusivity, and efficiency by enabling widespread participation and providing real-time data access. The integration of web-based platforms, interactive dashboards, Public Participation GIS (PPGIS), and social media offers various avenues for stakeholders to engage meaningfully and contribute to environmental decision-making.

Web-based platforms are increasingly crucial in modern EIAs. These digital tools allow stakeholders to access, view, and comment on EIA reports from remote locations. By providing an online portal where documents can be downloaded, reviewed, and annotated, these platforms ensure that information is accessible to a broad audience. This accessibility is particularly beneficial for stakeholders who may not have the means to attend physical meetings but still wish to voice their concerns or support for a project. For instance, stakeholders can upload their feedback through web portals, which EIA practitioners can then aggregate and analyze. This seamless integration of public input helps create more comprehensive and inclusive assessments, ensuring that diverse perspectives are considered.

Interactive dashboards provide another layer of engagement by offering visual representations of project progress and impact data. These dashboards enable stakeholders to track the development of an EIA in real time, making it easier to understand complex data through graphical formats such as charts, maps, and infographics. Such visualisation tools help demystify technical information, making it more digestible for non-experts. For instance, a dashboard might display air quality data over time, allowing users to see trends and anomalies at a glance. This capability empowers stakeholders to stay informed about the ongoing impacts of a project and to quickly identify areas of concern that require further investigation. Moreover, the ability to interact with the data—zooming into specific areas, filtering information by date, or comparing different metrics—enhances the user's analytical capabilities, fostering a deeper understanding of the environmental impacts being assessed.

Public Participation GIS (PPGIS) represents a significant advancement in engaging communities through localised data sharing and feedback. PPGIS platforms enable citizens to contribute geographically referenced information, often through user-friendly interfaces that resemble familiar mapping applications. For example, community members can mark points on a map indicating locations of environmental concern, such as areas affected by pollution or sites of historical significance. This localised input provides valuable context that might otherwise be overlooked in broader assessments. Research has shown that PPGIS can significantly enhance public participation and environmental awareness, leading to more informed decision-making processes (Mikhailova et al., 2012). By incorporating local knowledge into the EIA process, these tools ensure that assessments reflect the lived experiences and insights of the community, promoting greater buy-in and support for the final decisions.

Social media integration is another powerful tool for disseminating information and gathering public opinion on environmental projects. Platforms like Twitter, Facebook, and Instagram offer rapid and far-reaching channels for communicating EIA updates, findings, and opportunities for public engagement. Social media posts can include links to detailed reports, summaries of key findings, and invitations to virtual meetings or webinars. These platforms also facilitate two-way communication, allowing stakeholders to ask questions, express concerns, and provide feedback in real time. For instance, a project team might host a live Q&A session on Facebook to discuss the results of an air quality study, answering questions from the public as they arise. This immediacy fosters a sense of transparency and responsiveness that can be critical for building trust and credibility. Additionally, social media analytics can help EIA practitioners gauge public sentiment and identify common themes in the feedback received, informing subsequent stages of the assessment.

The integration of these technological tools not only enhances stakeholder engagement but also improves the overall quality of the EIA process. Digital platforms streamline document management and feedback collection, reducing administrative burdens and allowing practitioners to focus on substantive analysis. Interactive dashboards and PPGIS provide richer, more nuanced data sets that capture both the quantitative and qualitative dimensions of environmental impacts. Social media expands the reach of EIA communications, engaging a more diverse and potentially larger audience than traditional methods alone.

Insights and Implications

The chapter has explored the pivotal role of modern tools and technologies in enhancing Environmental Impact Assessments (EIAs). Geographic

Information Systems (GIS) enable robust data visualisation, spatial analysis, scenario planning, and stakeholder engagement, making complex data more accessible and decision-making more effective. Remote sensing techniques, including satellite imagery and aerial surveys, provide comprehensive environmental monitoring, while change detection methods highlight alterations over time. The integration of GIS with remote sensing further enhances data analysis, contributing to more informed EIAs.

Additionally, the chapter has delved into environmental modelling and data analytics as essential components of the EIA process. Predictive and integrated modelling frameworks support scenario analysis and uncertainty assessments, offering a comprehensive approach to environmental management. Data analytics, leveraging big data, real-time analysis, machine learning, and enhanced risk assessment, provides detailed insights and proactive measures for environmental protection. Together, these modern tools and technologies revolutionise the EIA process, supporting sustainable development and fostering transparent and inclusive decision-making.

Reference List

Ali, A. (2023, June 19). *The Role of Big Data in Environmental Conservation and Climate Change Mitigation* . Medium. https://asharibali.medium.com/the-role-of-big-data-in-environmental-conservation-and-climate-change-mitigation-5eb989914444

AuthorLastName, Initials. (2024). *Big data analytics and environmental performance: The moderating role of internationalization* . *Finance Research Letters* , 64, 105484. https://doi.org/10.1016/j.frl.2024.105484

Brovelli, M. A., Minghini, M., & Zamboni, G. (2015). *Public participation GIS: A FOSS architecture enabling field-data collection* . *International Journal of Digital Earth* , 8(5), 345-363. https://doi.org/10.1080/17538947.2014.887150

England, T. G. (2024, January 9). *Using GIS to Enhance Decision-Making* . ME&A. https://www.meandahq.com/using-gis-to-enhance-decision-making/

Imagery and Remote Sensing Software Integrated with GIS . (2022). Esri.com. https://www.esri.com/en-us/capabilities/imagery-remote-sensing/overview

People, G. I. S. (2023, September 24). *GIS for Environmental Impact Assessment* . GIS People:

https://www.gispeople.com.au/gis-for-environmental-impact-assessment/

Vorovencii, I. (2011). *Satellite remote sensing in environmental impact assessment: An overview* . *Unknown Journal* , 4(53). Retrieved from https://www.researchgate.net/publication/251484790_Satellite_Remote_Sensing_in_Environmental_Impact_Assessment_An_Overview

Werts, J. D., Mikhailova, E. A., Post, C. J., & Sharp, J. L. (2012). *An integrated WebGIS framework for volunteered geographic information and social media in soil and water conservation* . *Environmental Management* , 49(4), 816-832. https://doi.org/10.1007/s00267-012-9818-5

Chapter 4

Mitigation Measures and Strategies

Developing strategies to minimise adverse environmental impacts is a key component of Environmental Impact Assessments (EIA). Mitigation measures are essential for aligning development projects with the principles of sustainability, ensuring that environmental concerns are addressed proactively. The effort to integrate these strategies spans several domains, including design modifications, compensatory measures, innovative technologies, long-term monitoring plans, and sustainable architecture.

The chapter first explores how design adjustments can mitigate environmental impacts by incorporating sustainable materials, reusing existing structures, and tailoring projects to local ecosystems. It then delves into compensatory measures, such as habitat preservation agreements and in-lieu fee mitigation programmes, to restore lost environmental functions. Furthermore, the text examines the role of modern technologies, such as IoT devices and drones, in enhancing environmental assessment and monitoring. Long-term monitoring frameworks are also discussed, emphasising community involvement and the use of open data platforms. Finally, the chapter highlights sustainable architectural practices that promote resource conservation, energy efficiency, and alignment with circular economy principles.

Design Adjustments for Mitigation

Design modifications play a crucial role in mitigating the environmental impacts identified by Environmental Impact Assessments (EIA). By integrating sustainable design principles, employing existing structures, and tailoring projects to align with local ecosystems, we can significantly enhance biodiversity and resilience.

One of the fundamental strategies in sustainable design is the use of renewable materials. Employing materials with low embodied energy, recycled content, and non-toxic properties can dramatically reduce the ecological footprint of a project (Sustainable Design and Architecture | ArchitectureCourses.org, n.d.). For example, bamboo, a rapidly renewable resource, can be used for flooring instead of traditional hardwood, reducing deforestation and promoting sustainability. Additionally, recycled steel and concrete can replace virgin materials, minimising waste and conserving resources.

Modifying existing structures instead of opting for new constructions is another effective measure to reduce resource consumption. Adaptive reuse not only preserves historical and cultural sites but also reduces the demand for new materials and the energy required for construction. An excellent example of this is the transformation of old warehouses into residential lofts or commercial spaces. This approach can save substantial amounts of energy and raw materials while revitalising urban areas.

Tailoring projects to align with local ecosystems is essential for enhancing biodiversity and resilience. Developing site-specific designs that consider the local flora and fauna can create habitats that support wildlife. For instance, designing buildings with bird-friendly glass can prevent bird collisions, while incorporating bat boxes can provide nesting sites for bats. Additionally, creating corridors for wildlife movement within and around the project site can help maintain biodiversity and ecosystem functions.

Implementing green spaces and native plant species is another powerful strategy to enhance local biodiversity and ecosystem services. Green roofs and walls, for example, provide habitat for insects and birds, contribute to air purification, and help regulate building temperatures (Zhong et al., 2023). Using native plants in landscaping reduces the need for water, fertilisers, and pesticides, as these species are already adapted to the local climate and soil conditions. Furthermore, green spaces can act as community amenities, providing recreational areas and improving the overall quality of life for residents.

Employing sustainable design principles minimises the ecological footprint of a project. This involves considering the entire lifecycle of materials, from extraction to disposal. For example, choosing locally sourced materials reduces transportation emissions, while selecting products with longer lifespans decreases the frequency of replacements. Moreover, using non-toxic materials enhances indoor air quality, contributing to the health and well-being of occupants (Sustainable Design and Architecture | ArchitectureCourses.org, n.d.).

In addition to material selection, sustainable design encompasses various strategies to improve energy efficiency. Passive design techniques, such as optimising building orientation and incorporating natural ventilation, can significantly reduce the need for artificial heating and cooling. Similarly, implementing daylighting strategies, such as using skylights and light shelves, minimises reliance on electric lighting, thereby reducing energy consumption.

The adaptive reuse of structures is not only an environmentally responsible practice but also economically advantageous. Repurposing existing buildings can be more cost-effective than new construction, as it often requires less investment in infrastructure and utilities. Additionally, adaptive reuse projects can qualify for tax incentives and grants aimed at preserving historic buildings, further enhancing their financial viability. By conserving resources and reducing waste, adaptive reuse contributes to the overall sustainability of the built environment.

Location-specific adjustments are critical to aligning projects with local ecosystems. Understanding the unique characteristics of the project site allows for the development of tailored solutions that promote environmental stewardship. For instance, in coastal areas, designing buildings to withstand rising sea levels and storm surges can protect both the built and natural environments. In arid regions, incorporating water-saving technologies, such as xeriscaping and greywater recycling, can conserve valuable water resources.

Sustainable landscaping techniques play a vital role in enhancing local biodiversity and ecosystem services. Integrating green infrastructure, such as rain gardens and permeable pavements, helps manage stormwater runoff, reducing the risk of flooding and erosion. Additionally, planting native species supports local pollinators, such as bees and butterflies, which are essential for maintaining healthy ecosystems. By creating diverse and resilient landscapes, sustainable landscaping can mitigate the environmental impacts of development.

Compensatory Measures

Compensatory measures are critical in restoring or replacing lost environmental functions due to development activities. These measures ensure that even when development disrupts natural habitats, there is a proactive approach to mitigate such impacts and work towards sustainability. Here, we outline key compensatory strategies and their roles:

Engaging in environmental restoration projects, such as reforestation or wetland rehabilitation, is crucial in counterbalancing habitat loss. Reforestation involves planting trees in deforested areas to restore forest cover, which provides numerous benefits like carbon sequestration, soil stabilisation, and habitat for wildlife. Wetland rehabilitation focuses on restoring the hydrological and ecological functions of wetlands, which are essential for water purification, flood protection, and biodiversity support. According to Atkinson et al. (2022), restoration projects often face challenges due to local environmental gradients and successional dynamics. Nonetheless, long-term efforts tend to enhance overall biodiversity levels by gradually moving degraded sites closer to reference ecosystems.

Establishing habitat preservation agreements can provide credit for future project impacts. These agreements involve setting aside and managing land specifically for conservation purposes, ensuring that certain habitats remain untouched and protected from development. This strategy not only preserves existing ecosystems but also creates a buffer against future disturbances. It is essential to assess and plan these agreements carefully, considering historical land use and potential regional threats that might affect biodiversity outcomes. Effective preservation agreements can be tailored to specific species or habitat types, fostering a more targeted conservation effort.

Allowing developers to pay into a fund for conservation projects as compensation can streamline compliance with environmental regulations. Known as in-lieu fee mitigation, this approach provides an alternative to traditional onsite mitigation by pooling resources from multiple developers into a collective fund. This fund is then used to finance larger-scale, high-priority conservation projects that may be more effective than smaller, disjointed efforts. In-lieu fee mitigation options, along with strict oversight and open fund management, can lead to more strategic and effective conservation actions that focus on areas with high ecological value.

Initiatives focusing on specific species conservation are vital for preserving biodiversity functions. These programmes focus on the requirements of specific species whose habitats have suffered due to development, ensuring that they receive the protection and resources they need to survive. Conservation programmes may include breeding programmes, habitat restoration tailored to specific species, and protective measures against threats like poaching or invasive species. By concentrating efforts on vulnerable species, we can maintain the ecological roles they play, such as pollination, seed dispersal, and maintaining the balance of ecosystems.

To implement these compensatory measures effectively, it is necessary to follow certain guidelines. For habitat restoration projects, comprehensive planning and monitoring are essential. Restoration projects should consider the initial state of degradation and select appropriate interventions, such as replanting native species, managing invasive species, and restoring natural water flows. Success must be measured through long-term monitoring of biodiversity and ecosystem health indicators.

For conservation banking, it is important to establish clear criteria for bank credits and debits, ensuring transparency and accountability. Conservation banks should be strategically located to maximise ecological benefits, and ongoing management plans should be in place to maintain and enhance habitat quality over time. Engaging stakeholders, including local communities and environmental organisations, can also enhance the effectiveness and acceptance of conservation banking initiatives.

In-lieu fee mitigation requires robust governance structures to manage funds efficiently and transparently. Guidelines should outline priority areas for investment, selection criteria for conservation projects, and mechanisms for tracking progress and outcomes. Regular audits and public reporting can help maintain trust and demonstrate the tangible benefits of pooled funds.

Pollination and species habitat programmes must integrate scientific research and community involvement. Guidelines for these programmes should include identifying key species and their habitat requirements, creating comprehensive conservation plans, and implementing adaptive management strategies to respond to changing conditions. Community involvement is crucial, as local knowledge and participation can significantly bolster conservation efforts and ensure long-term success.

In all these compensatory measures, collaboration among developers, governments, conservation organisations, and local communities is pivotal. Effective environmental restoration and compensation require shared goals, coordinated actions, and sustained commitment. By following these guidelines and fostering partnerships, we can better restore and maintain the environmental functions crucial for sustainable development.

Innovative Technological Solutions

Modern technologies are increasingly becoming a cornerstone in efforts to mitigate environmental impacts. This section delves into how various advanced technologies can help manage and reduce adverse environmental effects effectively.

Implementing IoT devices can provide real-time data on environmental conditions. The Internet of Things (IoT) refers to the network of physical devices embedded with sensors, software, and other technologies that connect and exchange data over the internet. In the context of environmental impact mitigation, IoT devices can monitor air quality, water quality, soil conditions, and even noise levels in real-time. These devices offer numerous advantages, including the ability to detect changes instantly and initiate corrective measures swiftly. For instance, smart sensors placed in industrial areas can immediately alert authorities to dangerous levels of pollutants, enabling rapid response to mitigate potential damage. Moreover, continuous data collection enhances the accuracy of environmental assessments, making it easier for stakeholders to make informed decisions.

Utilising drones for environmental assessment and monitoring can enhance efficiency. Drones, also known as unmanned aerial vehicles (UAVs), have transformed environmental monitoring by providing comprehensive aerial views that were previously difficult to obtain. Equipped with high-resolution cameras and various sensors, drones can capture detailed images and data from hard-to-reach areas such as dense forests, wetlands, and mountainous regions. This technology is particularly useful for monitoring deforestation, surveying wildlife habitats, and assessing the health of ecosystems. Drones can cover large areas quickly and repeatedly, offering time-series data that helps track environmental changes over time. Additionally, they can be deployed in hazardous environments where human presence might be risky, ensuring safety while gathering essential information. This makes them invaluable tools for enhancing the efficiency and accuracy of environmental assessments.

Employing natural organisms to remove pollutants can offer an eco-friendly mitigation option. Bioremediation techniques use microorganisms, plants, or fungi to detoxify contaminated environments naturally. This method is highly effective in treating polluted soil and water by breaking down harmful substances into less toxic forms. One notable example is the use of bacteria to clean up oil spills. These microorganisms consume hydrocarbons, thereby reducing pollution. Similarly, certain plants, referred to as hyperaccumulators, can absorb heavy metals from contaminated soils, which are subsequently harvested and safely disposed of. The advantage of bioremediation lies in its sustainability and minimal ecological footprint, unlike conventional chemical treatments that may introduce additional pollutants. Further research and development in this field can broaden the range of pollutants that bioremediation can address, making it a versatile tool for environmental management.

Adopting energy-efficient technologies in project design can significantly reduce environmental impacts. Energy consumption is a major contributor to environmental degradation, primarily through the emission of greenhouse gases from fossil fuel use. Integrating energy-efficient technologies in the initial stages of project design can lead to substantial reductions in a project's overall environmental footprint. This encompasses a wide range of practices, from using LED lighting and energy-efficient HVAC systems to incorporating renewable energy sources like solar and wind power. Buildings designed with energy efficiency in mind often include features such as high-performance insulation, optimised natural lighting, and advanced energy management systems. These innovations not only lower operational costs but also contribute to broader efforts to combat climate change by reducing carbon emissions. For instance, smart grid technologies facilitate efficient electricity distribution and consumption, further enhancing environmental benefits.

Long-Term Monitoring Plans

Establishing long-term monitoring frameworks is crucial for the sustainability of environmental mitigation efforts. By implementing these frameworks, we can ensure that measures taken to mitigate adverse environmental impacts remain effective over time. One fundamental aspect of this approach is establishing clear metrics for tracking ecosystem health post-development.

Clear metrics provide a structured way to assess whether environments are recovering or deteriorating. These metrics might include biodiversity indices, water and soil quality parameters, and the presence of indicator species. For example, measuring the population diversity of native plants and animals before and after a developmental project can reveal impacts on local biodiversity. Regular monitoring using these metrics allows stakeholders to identify trends and intervene when necessary.

Another essential element is involving local communities in the monitoring process.

Community-based monitoring (CBM) empowers residents to take an active role in safeguarding their environment. CBM not only enhances data collection through citizen science but also promotes stewardship and collective responsibility. When community members participate in monitoring, they gain a deeper understanding of their environment and become more invested in its preservation. This involvement can significantly enhance the capacity for early detection of environmental changes, enabling quicker responses to emerging issues. According to Muhamad Khair et al. (2021), CBM encourages stakeholder participation, which greatly impacts environmental sustainability.

To further improve transparency and accessibility, utilising open data platforms for sharing monitoring results is imperative. Open data ensures that information about the health of ecosystems is readily available to all stakeholders, including policymakers, scientists, and the public. These platforms enable real-time data sharing and analysis, fostering a culture of accountability and informed decision-making. Local authorities and community organizations, for instance, can access an online portal that lists the parameters of a river's water quality, facilitating cooperative efforts to address any detected pollutants.

Moreover, developing a responsive monitoring framework is key to addressing evolving environmental challenges. A responsive system allows for adaptive management, meaning that mitigation strategies can be adjusted based on new data and changing conditions. This flexibility is particularly important given the dynamic nature of ecosystems and the potential for unanticipated impacts from development projects. For instance, if initial monitoring reveals that a wetland restoration project is not improving water quality as expected, adjustments such as altering plant species or modifying hydrology can be made promptly to achieve desired outcomes.

The integration of advanced technologies can further enhance the effectiveness of monitoring frameworks. Using IoT devices and remote sensing technology provides detailed, continuous data streams on various environmental parameters. These technologies facilitate the collection of high-resolution data across extensive geographical areas, often in real-time. The deployment of sensor networks in forest ecosystems, for example, can monitor soil moisture levels, temperature fluctuations, and tree health, providing comprehensive insights into ecosystem dynamics. Combining this data with machine learning algorithms can offer predictive analytics, helping anticipate future environmental conditions and plan accordingly.

Implementing training programmes for local communities and stakeholders on the use of monitoring tools and data interpretation is also vital. These programmes ensure that participants have the necessary skills to effectively gather and analyse data, thereby maximising the benefits of CBM initiatives. Providing workshops and educational resources fosters capacity-building and reinforces the long-term sustainability of monitoring efforts.

Additionally, creating partnerships between governmental bodies, non-governmental organisations (NGOs), academic institutions, and local communities enriches the monitoring process. Such collaborations bring together diverse expertise, resources, and perspectives, enhancing the overall quality of environmental assessments. Universities, for instance, can contribute scientific knowledge and technical support, while NGOs often have on-the-ground experience and community trust. Government agencies can provide regulatory frameworks and policy support, ensuring that monitoring efforts align with broader environmental goals.

Financial sustainability is another critical aspect to consider. Long-term funding mechanisms must be established to support ongoing monitoring activities. This can include securing grants, developing endowment funds, or integrating monitoring costs into project budgets during the planning phase. Ensuring consistent financial backing prevents gaps in data collection and allows for the continuous evaluation of mitigation measures.

Finally, regular reporting and communication of monitoring findings are essential to maintaining stakeholder engagement and driving continuous improvement. Clear, well-structured reports that highlight key findings, trends, and recommended actions should be disseminated to all relevant parties. This transparency builds trust and encourages collaborative problem-solving. For example, quarterly reports on coastal erosion rates and habitat changes can guide local governments in implementing timely protective measures.

Integration of Sustainable Architecture

Sustainable architectural practices play a crucial role in reducing environmental impacts, particularly within the framework of Environmental Impact Assessments (EIA). These practices ensure that buildings and structures not only meet present needs but also do so without compromising the ability of future generations to meet theirs. By integrating resource conservation and energy efficiency into the design phase, architects and builders can create more sustainable projects that align with the broader goals of environmental stewardship.

Promoting Resource Conservation and Energy Efficiency

In the design phase, promoting resource conservation and energy efficiency is paramount. Natural resources are finite, and their depletion can lead to severe environmental consequences. By incorporating advanced insulation, efficient HVAC systems, and the use of renewable energy sources like solar panels, architects can significantly reduce a building's energy consumption. This approach not only minimises the carbon footprint but also results in substantial cost savings over the building's lifecycle.

Encouraging Eco-Friendly Construction Techniques

Eco-friendly construction techniques involve using materials and methods that have minimal environmental impact. For instance, sourcing materials locally can reduce transportation emissions, while selecting recycled or sustainably harvested materials helps lower the ecological footprint. Innovative construction techniques, such as modular buildings, can also contribute to sustainability by minimising material waste and reducing construction time.

One notable example is the use of cross-laminated timber (CLT), a renewable material that has gained popularity for its strength and low environmental impact. CLT buildings sequester carbon, meaning they can help offset emissions from other parts of the project. Encouraging these practices among practitioners ensures that sustainable principles become a standard part of construction processes.

Reducing Waste During the Project Lifecycle

Projects generate significant amounts of waste throughout their lifecycle—from construction through demolition or repurposing. Sustainable architectural practices aim to minimise this waste through careful planning and implementation. During construction, strategies such as precise material calculations, robust recycling programmes, and onsite waste management plans are essential. Additionally, designing buildings for adaptability and eventual deconstruction rather than demolition can further support waste reduction efforts.

A fundamental aspect of waste reduction involves using durable materials that extend the building's lifespan. This longevity means fewer resources are needed to maintain or replace the structure over time. Moreover, integrating technologies that enable easy upgrades can adapt buildings to new uses without substantial renovation work, thus conserving resources and reducing waste.

Aligning with Circular Economy Principles

The concept of a circular economy focuses on maintaining the value of products, materials, and resources in the economy for as long as possible. It contrasts with the traditional linear economy, which follows a "take-make-dispose" model. Sustainable architecture aligns perfectly with circular economy principles by prioritising designs that facilitate reuse, refurbishment, and recycling.

For example, modular components designed for easy disassembly can be reused in other projects, thus extending their lifecycle. The use of reclaimed materials, such as bricks from demolished buildings or repurposed steel, supports the circular flow of resources. Additionally, employing design for deconstruction—where buildings are constructed with future disassembly in mind—ensures that components can be efficiently repurposed at the end of their useful life.

Promoting Resource Conservation and Energy Efficiency in the Design Phase

Resource conservation and energy efficiency should be integral from the initial conceptual stages of a project. Architects must consider various factors, such as site orientation, natural lighting, and the local climate, to optimise energy use. Incorporating passive solar design, which leverages sunlight for heating and daylighting, reduces reliance on artificial lighting and heating systems.

Effective thermal insulation is critical for maintaining indoor comfort while decreasing energy demands. High-performance windows and doors, along with airtight construction, help prevent heat loss in winter and keep interiors cool in summer. Additionally, integrating renewable energy systems like photovoltaic panels or geothermal heating can make buildings self-sustaining in terms of energy needs.

Encouraging Adoption of Eco-Friendly Construction Techniques Among Practitioners

Practitioners in the building industry must adopt eco-friendly construction techniques to promote sustainability. Training and awareness programmes can equip them with the knowledge needed to implement these methods effectively. Certifications and standards, such as LEED (Leadership in Energy and Environmental Design) or BREEAM (Building Research Establishment Environmental Assessment Method), provide guidelines and benchmarks for constructing green buildings.

Moreover, collaboration among stakeholders—including architects, engineers, contractors, and clients—is essential for successful implementation. By working together, they can ensure that sustainable practices are integrated seamlessly into all stages of the project. Policies and incentives from government agencies can further support the adoption of eco-friendly techniques on a broader scale.

Contributing to the Reduction of Waste During the Project's Lifecycle and Supporting Sustainability

Managing waste effectively throughout a project's lifecycle is key to sustainability. During the construction phase, accurate material estimation prevents excess ordering, and establishing recycling stations onsite allows materials such as metals, plastics, and wood to be diverted from landfills. Efficient logistics planning ensures that materials arrive as needed, reducing the likelihood of damage and waste.

In the operational phase, buildings designed for maintenance ease and adaptability experience less wear and tear, resulting in fewer renovations and replacements. Implementing systems for water conservation, such as low-flow fixtures and rainwater harvesting, reduces water waste and promotes resource efficiency.

Aligning with Circular Economy Principles That Focus on Resource Efficiency

Circular economy principles can be applied effectively to building design and construction. One practical approach is to design for longevity and flexibility, ensuring that structures can adapt to changing needs over time. Buildings that can be easily modified or expanded without significant structural changes support resource efficiency.

Another principle is prioritising the use of materials with high recycled content and those that can be readily reused or recycled. For instance, using reclaimed wood or recycled metal reduces the demand for virgin materials and decreases the project's overall environmental impact. Designing for waste minimization from the outset—such as through prefabrication or modular construction—also aligns with circular economy ideals by reducing offcuts and surplus materials.

Final Thoughts

Developing strategies to minimise adverse environmental impacts is essential in EIA, supporting the overall goal of sustainable project outcomes. This chapter has outlined various approaches, such as design adjustments for mitigation, compensatory measures, innovative technological solutions, long-term monitoring plans, and integrating sustainable architecture. Emphasising the importance of employing renewable materials and modifying existing structures, it highlighted how these practices can reduce resource consumption and enhance biodiversity. Additionally, aligning projects with local ecosystems and incorporating energy-efficient technologies further contribute to minimising ecological footprints and promoting sustainability.

The chapter also discussed the role of compensatory measures and technological innovations in mitigating environmental impacts. Techniques like bioremediation and the use of IoT devices for real-time monitoring have shown promise in managing environmental conditions more effectively. Long-term monitoring frameworks ensure that mitigation efforts remain effective over time, involving local communities and utilising open data platforms. By adopting sustainable architectural practices and aligning them with circular economy principles, we can significantly reduce waste and resource consumption throughout a project's lifecycle. Implementing these strategies collectively supports the overarching goal of achieving sustainable development while minimising adverse environmental impacts.

Reference List

7 Concept Of Green Buildings: A Complete Guide To Sustainable Architecture | Archiroots . (2023, November 25). https://archiroots.com/concept-of-green-buildings-a-complete-guide/

Atkinson, J., Brudvig, L. A., Mallen-Cooper, M., Nakagawa, S., Moles, A. T., & Bonser, S. P. (2022, May 12). *Terrestrial ecosystem restoration increases biodiversity and reduces its variability, but not to reference levels: A global meta-analysis* (T. Crowther, Ed.). Ecology Letters. https://doi.org/10.1111/ele.14025

Document Display (PURL) | NSCEP | US EPA . (2024). Epa.gov. https://nepis.epa.gov/Exe/ZyPURL.cgi?Dockey=P100EUJG.TXT

Jobya. (2024, April 26). *Environmental: The Role of Technology in Shaping Environmental Careers* . Jobya. https://jobya.com/library/industries/environmental/articles/the_role_of_technology_in_shaping_environmental_careers

Mahajan, S. (2022, August 24). *Design and development of an open-source framework for citizen-centric environmental monitoring and data analysis* . Scientific Reports. https://doi.org/10.1038/s41598-022-18700-z

Muhamad Khair, N. K., Lee, K. E., & Mokhtar, M. (2021, July 1). *Community-based monitoring for environmental sustainability: A review of characteristics and the synthesis of criteria* . Journal of Environmental Management. https://doi.org/10.1016/j.jenvman.2021.112491

Oloruntobi, O., Mokhtar, K., Mohd Rozar, N., Gohari, A., Asif, S., & Chuah, L. F. (2023, April). *Effective technologies and practices for reducing pollution in warehouses - A review* . Cleaner Engineering and Technology. https://doi.org/10.1016/j.clet.2023.100622

Sustainable Design and Architecture | ArchitectureCourses.org . (n.d.). Www.architecturecourses.org. https://www.architecturecourses.org/learn/sustainable-design-and-architecture

Zhong, W., Schroeder, T., & Bekkering, J. (2023, April 17). *Designing with nature: Advancing three-dimensional green spaces in architecture through frameworks for biophilic design and sustainability*

. Frontiers of Architectural Research. https://doi.org/10.1016/j.foar.2023.03.001

jenks2026. (2024, January 30). *Sustainability and Green Building Practices* . Green.org. https://green.org/2024/01/30/sustainability-and-green-building-practices/

Chapter 5

Public Consultation and Stakeholder Engagement

Public consultation and stakeholder engagement are fundamental to the Environmental Impact Assessment (EIA) process. Effective engagement with stakeholders ensures that a diverse range of perspectives is considered, leading to more informed and balanced decision-making in resource management. By incorporating public opinion, environmental professionals can tailor their assessments to reflect community needs and concerns, thereby enhancing the legitimacy and acceptance of proposed projects.

In this chapter, various methods of public engagement will be explored, including public meetings, surveys, focus groups, and online platforms. Each method offers unique advantages and is suited to different contexts and stakeholder groups. The chapter will also discuss best practices for organising these engagement activities, such as clear communication, providing background materials, and ensuring inclusivity. Additionally, real-world case studies will illustrate the benefits of stakeholder engagement in achieving sustainable outcomes. Lastly, the chapter will address common challenges in public consultations and offer strategies to overcome them, ensuring a comprehensive understanding of effective public engagement in the EIA process.

Methods of Public Engagement

Engaging the public in the Environmental Impact Assessment (EIA) process is essential for gathering diverse viewpoints and making more informed decisions. Public participation ensures that different perspectives are considered, leading to more effective and sustainable outcomes.

Public meetings are a traditional but highly effective approach to engaging stakeholders in the EIA process. These organised events provide a forum where community members, local organisations, and other interested parties can come together to discuss the proposed project. Public meetings facilitate face-to-face interaction between the project proponents and the stakeholders, allowing for an open exchange of information and ideas. This direct engagement helps build trust and ensures that stakeholder concerns are heard and addressed. Public meetings can be structured as town halls, workshops, or seminars, depending on the complexity of the project and the level of interest among the stakeholders.

To ensure productive public meetings, it is crucial to adopt specific guidelines. Clear communication about the meeting's purpose and agenda helps set expectations. Providing background materials in advance, such as project descriptions and potential impacts, allows participants to come prepared with informed questions and comments. It is also beneficial to designate a neutral facilitator who can manage the discussion, ensuring that all voices are heard and that the conversation remains on topic.

Another effective method for engaging the public is through surveys and questionnaires. These tools are valuable for collecting quantitative data on community attitudes and opinions regarding the proposed project. Surveys and questionnaires can reach a broad audience, including those who may not be able to attend public meetings. They offer a convenient way for stakeholders to provide input at their own pace and convenience.

Designing effective surveys requires careful consideration of several factors. Questions should be clear, concise, and unbiased to avoid skewing the responses. Including a mix of closed-ended and open-ended questions can capture both quantitative data and qualitative insights. Closed-ended questions, such as multiple-choice or Likert-scale items, make it easy to analyse trends and patterns in the data. Open-ended questions, on the other hand, allow respondents to express their thoughts and concerns in their own words, providing deeper insights into specific issues.

Focus groups are another valuable tool for public engagement in the EIA process. These small, guided discussions bring together selected community members to explore particular aspects of the project in depth. A skilled facilitator typically moderates focus groups, directing the discussion and making sure that everyone has a chance to contribute.

The intimate setting of focus groups allows for a more relaxed and detailed exploration of stakeholder concerns and suggestions. Participants can engage in meaningful dialogue, share personal experiences, and brainstorm potential solutions. To maximise the effectiveness of focus groups, it is important to select a diverse group of participants representing different interests and perspectives within the community. The facilitator's role is critical in creating an inclusive and respectful environment where everyone feels comfortable sharing their views.

As technology continues to evolve, online platforms have become increasingly popular for engaging stakeholders in the EIA process. Digital tools and social media offer innovative ways to reach a larger and more diverse audience, including those who may be unable to attend in-person meetings due to geographical or logistical constraints. Online engagement can take various forms, including virtual town halls, webinars, discussion forums, and social media campaigns.

Utilising online platforms requires a strategic approach to ensure effective communication and participation. Project proponents should create user-friendly websites or portals where stakeholders can access information about the project, submit feedback, and participate in discussions. Social media channels can be used to share updates, promote engagement opportunities, and foster ongoing dialogue with the community. Virtual town halls and webinars should be scheduled at convenient times and recorded for those who cannot join live. Additionally, ensuring accessibility for individuals with disabilities and providing multilingual support can help broaden participation.

Examples of Community Involvement

To showcase the pivotal role of community involvement in Environmental Impact Assessments (EIAs), we turn to several real-world case studies. These instances underscore how integrating stakeholder engagement strategies can lead to more informed and effective resource management decisions.

Case Study 1: Coastal Restoration Project
In a coastal restoration project, local stakeholders played a crucial role in identifying key areas for restoration that experts might have overlooked. This project aimed to restore natural habitats along the coastline to mitigate erosion and promote biodiversity. While environmental scientists brought their technical expertise to the table, it was the input from local fishermen, residents, and small

business owners that highlighted specific zones requiring urgent attention. These stakeholders, with their intimate knowledge of the area, could identify places where erosion was most severe or where species loss had been particularly noticeable. Their insights led to the prioritisation of these critical areas, ensuring that the restoration efforts were both targeted and effective. The outcome was a more robust and resilient coastline that better served the ecological and economic needs of the community.

Case Study 2: Urban Development

Direct community input on transportation needs has also demonstrated significant benefits in urban development projects. In a metropolitan area seeking to enhance its public transit system, the city planners initially focused on expanding routes to affluent neighbourhoods. However, stakeholder engagement sessions revealed a different set of priorities among under-represented communities. Residents from these areas voiced concerns about the lack of accessible and reliable transportation options, which hindered their daily commutes and access to essential services. In response, the city planners redesigned certain routes to better serve these neighbourhoods. This adjustment not only addressed the immediate transportation needs of these communities but also fostered greater inclusivity and equity within the urban planning process. The redesign resulted in a more comprehensive transit system that improved connectivity and quality of life for all residents.

Case Study 3: Renewable Energy Projects

Early public engagement in renewable energy projects can establish credibility and mitigate opposition. Consider a wind farm development proposed in a rural area. Initially met with resistance due to concerns about noise, visual impacts, and potential effects on local wildlife, the project developers decided to engage the community early in the planning process. They organised public meetings, workshops, and site visits to address these concerns transparently. By involving community members in the decision-making process, the developers were able to build trust and demonstrate their commitment to addressing stakeholder concerns. They implemented measures such as altering the siting of turbines to minimise visual impact and introducing noise reduction technologies. Furthermore, they assured the community that they would monitor and mitigate any adverse effects on wildlife. As a result, the project gained widespread acceptance and proceeded with minimal opposition. This case exemplifies how proactive engagement can transform potential conflict into collaboration, paving the way for successful project implementation.

Case Study 4: Pollution Control Initiatives

Engaging affected communities has proven vital in developing practical and effective pollution control initiatives. In an industrial region grappling with air and water pollution, local residents formed a coalition to voice their health and environmental concerns. Their firsthand experiences provided invaluable data on pollution hotspots and the types of pollutants affecting their daily lives. Collaborating with environmental experts and policymakers, the community developed targeted mitigation strategies. These included installing air quality monitoring stations, enhancing wastewater treatment processes, and introducing stricter emissions regulations for nearby factories. By incorporating local knowledge and prioritising community well-being, the initiative achieved significant improvements in environmental quality. The community's active participation ensured that the solutions were not only scientifically sound but also socially acceptable and sustainable.

Each of these case studies highlights the tangible benefits of involving stakeholders in the EIA process. Local knowledge and insight can significantly enhance the accuracy and relevance of environmental assessments, leading to more effective and equitable outcomes. By valuing and incorporating community input, project developers and policymakers can overcome challenges, build trust, and achieve long-term sustainability.

Challenges in Stakeholder Consultations

Public consultation and stakeholder engagement are pivotal to the Environmental Impact Assessment (EIA) process, ensuring that diverse viewpoints are considered in decision-making. However, several obstacles can hinder effective public consultations and stakeholder engagement. This section will identify and discuss these common obstacles, highlighting their impact on the EIA process.

A significant challenge in public consultations is limited awareness and understanding of the EIA process among stakeholders. Many community members and other stakeholders may be unaware of what an EIA entails, its objectives, and how it impacts their environment. This lack of knowledge can lead to disinterest or passive participation, which diminishes the quality of feedback received during consultations. When stakeholders do not fully grasp the significance of the EIA, they may not provide meaningful input, missing opportunities to influence the project outcomes positively.

To mitigate this issue, educational initiatives should be undertaken prior to and during the public consultation process. Stakeholders need access to clear, concise information about the EIA process, including case studies, visual aids, and explanatory sessions. By enhancing their understanding, stakeholders are more likely to engage actively and provide insightful contributions that can shape better environmental management practices.

Diverse interests and conflicts among stakeholders present another substantial obstacle. Public consultations often bring together individuals and groups with varying, and sometimes opposing, interests. For example, a proposed infrastructure project might receive support from business owners anticipating economic growth, while environmentalists and local residents might oppose it due to potential ecological damage and disruption to their livelihoods. Such conflicting interests can complicate the engagement process, making it challenging to reach a consensus.

Effective facilitation and conflict resolution strategies are essential to addressing these differences. Facilitators should ensure that all voices are heard and respected, creating a balanced platform where stakeholders can express their concerns and aspirations. Techniques such as mediation and negotiation can help bridge gaps between conflicting parties, fostering a more collaborative atmosphere. By acknowledging and addressing diverse interests, the EIA process can integrate a broader range of perspectives, leading to more comprehensive and accepted outcomes.

Resource constraints are another barrier that can compromise the effectiveness of stakeholder engagement activities. Conducting thorough and inclusive consultations requires adequate funding, time, and human resources. Limited financial resources can restrict the scope of engagement activities, reducing the number of meetings, workshops, and outreach efforts. Similarly, tight project timelines might pressure consultation processes, leading to hurried discussions that fail to capture the depth of stakeholder concerns.

Overcoming resource constraints involves strategic planning and allocation of available resources. Prioritising key stakeholder groups and leveraging existing community networks can enhance outreach without incurring significant costs. Additionally, engaging experts in volunteer capacities or seeking partnerships with organisations invested in public participation can supplement resource limitations. Ensuring sufficient time for consultation processes is crucial; well-scheduled engagements allow for richer dialogues and more thoughtful input from stakeholders.

Technological barriers also pose challenges in public consultations and stakeholder engagement. While technology offers innovative ways to enhance participation, such as virtual meetings, online surveys, and interactive platforms, it can inadvertently create barriers for less tech-savvy community members. Older adults, individuals in rural areas with limited internet access, and those unfamiliar with digital tools may find it difficult to participate effectively through technological means.

Addressing technological barriers requires a blended approach that combines traditional and modern engagement methods. In-person meetings, printed materials, community notice boards, and radio announcements can complement digital tools, ensuring inclusivity. Additionally, providing technical assistance and training to stakeholders can empower them to use technological platforms confidently. By adopting a multimodal strategy, the EIA process can maximise participation across diverse demographic groups, ensuring that all voices are heard irrespective of their technological proficiency.

Benefits of Inclusive Decision-Making

In the realm of Environmental Impact Assessment (EIA), inclusive stakeholder engagement is a cornerstone of success. By actively involving stakeholders in the EIA process, projects not only gain valuable insights but also foster trust and support within the community. This subpoint details the significant advantages that inclusive stakeholder engagement brings to the EIA process, elucidating how it enhances project acceptance, improves design and implementation, leads to sustainable outcomes, and empowers communities.

Enhanced Project Acceptance

One of the foremost benefits of inclusive stakeholder engagement is enhanced project acceptance. When community members are involved in decision-making from the onset, they are more likely to understand the project's objectives, processes, and potential impacts. This understanding translates into greater public support. For instance, consider a proposed industrial facility. Without stakeholder engagement, the local community might perceive the project as a threat to their environment and health. However, if stakeholders are consulted and their concerns addressed early on, they can become advocates rather than adversaries. They may even provide unique historical or cultural insights that ensure the project aligns with community values.

Furthermore, transparency in communication fosters trust. When stakeholders see that their input genuinely influences decisions, their scepticism diminishes. A transparent consultation process allows stakeholders to voice their concerns and see how these concerns are mitigated in the project plan. This participatory approach builds a sense of ownership among community members, making them more likely to support the project's implementation and ongoing operations.

Improved Project Design and Implementation

Another critical advantage of inclusive stakeholder engagement is the potential for improved project design and implementation. Stakeholders often bring diverse perspectives and expertise that can lead to innovative solutions and practical adaptations in project plans. For example, local residents might suggest modifications based on their intimate knowledge of the area's geography, climate, and biodiversity. These suggestions can help optimise the project's design to better suit the local context.

Consider a scenario where a new transportation infrastructure project is underway. Local commuters might highlight specific traffic patterns or congestion points that external consultants may overlook. By incorporating these insights, planners can design a more efficient and responsive transportation system. Similarly, indigenous communities might share traditional ecological knowledge that helps safeguard critical habitats and species during the project's execution.

Moreover, stakeholder engagement can preemptively identify potential issues and conflicts before they escalate. Early detection of these challenges allows project managers to implement corrective measures, minimising delays and cost overruns. Inclusive engagement creates a platform for continuous feedback, ensuring that the project evolves in response to real-time observations and recommendations.

Sustainable Outcomes

Projects designed with community input are inherently more sustainable. Sustainability, in this context, refers to the ability of a project to provide long-term environmental, social, and economic benefits. When stakeholders are involved in the planning process, their contributions often reflect a deep commitment to preserving and enhancing their surroundings. This input is invaluable in crafting solutions that balance development goals with environmental stewardship.

For instance, a renewable energy project that engages local farmers and landowners can lead to mutually beneficial arrangements, such as integrating wind turbines into agricultural landscapes. The stakeholders' willingness to host these installations on their land ensures the project's feasibility while promoting sustainable energy production. Furthermore, community-driven initiatives often prioritise resource conservation and ecosystem restoration, leading to positive environmental outcomes.

Sustainable outcomes extend beyond environmental considerations. Projects that consider social dynamics and local needs are more likely to succeed in the long run. For example, a community-led water management project can address both technical and social aspects of water use, ensuring equitable distribution and long-term viability. By aligning the project with local priorities, stakeholders are motivated to maintain and support it over time, reducing the risk of abandonment or neglect.

Empowered Communities

Inclusive decision-making also empowers communities, enhancing their capacity to engage in future initiatives. When stakeholders are involved in the EIA process, they gain knowledge and skills that enable them to participate more effectively in subsequent projects. This empowerment can manifest in several ways, from increased civic participation to advocacy for better policies and practices.

Community empowerment is particularly significant in contexts where marginalised or under-represented groups have historically been excluded from decision-making processes.

Involving these groups not only rectifies past injustices but also enriches the project with diverse voices and perspectives. For instance, women in rural communities often play pivotal roles in managing natural resources. Their inclusion in stakeholder consultations can lead to more equitable and effective resource management strategies.

Additionally, empowered communities are better equipped to hold project developers and authorities accountable. They can advocate for compliance with environmental regulations, monitor project impacts, and ensure that commitments made during the consultation process are fulfilled. This watchdog role contributes to greater transparency and accountability in project governance.

Moreover, inclusive engagement fosters social cohesion and resilience. When communities collaborate on shared goals, they build stronger relationships and networks. These connections can be invaluable in times of crisis, enabling collective action and mutual support. For example, communities that have worked together on an EIA project may be better prepared to respond to environmental emergencies, such as natural disasters or pollution incidents.

Effective Facilitation of Public Engagement

Public consultation and stakeholder engagement are essential components of the Environmental Impact Assessment (EIA) process. Engaging the public effectively ensures that their opinions, concerns, and insights are considered in decision-making, leading to more informed decisions and better resource management. In this section, we will explore measures that ensure productive and inclusive public engagement processes.

Transparency and Inclusivity are fundamental to successful stakeholder engagement. Structured public meetings play a critical role in fostering open dialogue by clearly communicating project details, objectives, and potential impacts. These meetings should be well-publicized and accessible to all members of the community, including those with disabilities or language barriers. Providing detailed and understandable information can help demystify the EIA process, making it easier for stakeholders to engage meaningfully. Additionally, employing a variety of communication methods—such as visual aids, presentations, and Q&A sessions—can enhance comprehension and encourage participation from diverse groups.

Anonymity and honest feedback are crucial in gathering authentic input from stakeholders. Surveys and questionnaires can be highly effective tools in this regard, allowing individuals to express their concerns and suggestions without fear of retribution or judgement. Ensuring anonymity encourages more candid responses, which can provide valuable insights into the community's true sentiments about a project. To maximise the effectiveness of these tools, questions should be designed to be clear and unbiased, covering various aspects of the project. Regularly analysing and acting on survey results demonstrates to the community that their feedback is valued and taken seriously, further building trust in the engagement process.

In-depth discussions through focus groups offer another layer of engagement, enabling a deeper understanding of specific community issues. Focus groups consist of small, moderated sessions where selected participants can discuss their views in detail. This setting allows for a more thorough exploration of topics that might not be fully addressed in larger public meetings or surveys. Moderators play a vital role in guiding the conversation, ensuring that all voices are heard and that discussions remain focused and productive. By selecting a diverse cross-section of the community for these groups, organisers can gain a more comprehensive understanding of the varying perspectives and concerns related to the project.

Continuous dialogue through online platforms has become increasingly important in ensuring ongoing interaction with the community. Digital tools such as websites, social media, and dedicated forums provide a convenient way for stakeholders to stay informed and engaged. These platforms allow community members to voice their opinions and ask questions at their convenience, making engagement more accessible to those who cannot attend in-person events. Additionally, online platforms can facilitate real-time updates and feedback, helping to sustain an ongoing conversation between project developers and the community. Ensuring that these digital spaces are user-friendly and regularly monitored can enhance the overall effectiveness of public engagement efforts.

By combining these measures, practitioners can create a robust and inclusive public engagement process. Transparency and inclusivity lay the foundation for open communication, while anonymity in feedback mechanisms fosters honesty. In-depth discussions through focus groups provide detailed insights, and continuous dialogue via online platforms ensures ongoing participation. Together, these strategies contribute to a more participatory EIA process, ultimately leading to more informed decision-making and better outcomes for both the environment and the community.

As environmental professionals and practitioners, it is essential to recognise the value of incorporating these engagement measures into the EIA process. Not only do they enhance the quality and legitimacy of the assessments, but they also build stronger relationships with the communities affected by proposed projects. For policymakers and regulators, understanding these engagement strategies can inform the development of more effective environmental policies and guidelines. Lastly, students and academics can benefit from studying these approaches, gaining practical insights that bridge theory and application in the fields of environmental science and resource management.

In practice, implementing these engagement measures requires careful planning, coordination, and a commitment to genuinely valuing stakeholder input. Public meetings should be scheduled at convenient times and locations, and efforts should be made to reach out to traditionally under-represented groups. Surveys and questionnaires need to be crafted thoughtfully to elicit meaningful responses, and focus groups should be facilitated in a manner that encourages open and respectful dialogue. Online platforms must be maintained and actively used to keep the community informed and engaged throughout the project's lifecycle.

Ultimately, the goal of these public engagement measures is to create a more democratic and equitable EIA process. By ensuring that all voices are heard and considered, we can achieve more sustainable and acceptable outcomes for development projects. This inclusive approach not only improves the quality of environmental assessments but also enhances the legitimacy and acceptance of the final decisions.

Final Thoughts

The chapter has highlighted various methods of public engagement in the Environmental Impact

Assessment (EIA) process, emphasising their significance in ensuring informed and effective decision-making. It discussed traditional approaches such as public meetings, surveys, and focus groups, as well as modern tools like online platforms. Each method provides unique advantages that contribute to comprehensive stakeholder involvement, enhancing the overall integrity and acceptance of the EIA process. By incorporating diverse perspectives, projects can better address community concerns and achieve more sustainable outcomes.

These public engagement strategies are crucial for fostering trust and collaboration between project developers and stakeholders. Effective facilitation through clear communication, conflict resolution, and inclusive practices ensures that all voices are heard and valued. The chapter's insights into real-world case studies demonstrate how proactive stakeholder involvement can lead to successful project implementation and long-term community benefits. As environmental professionals, policymakers, and students consider these methods, they can improve the quality and legitimacy of environmental assessments, ultimately contributing to more equitable and sustainable resource management.

Chaper 6

Environmental Components in EIA

An Environmental Impact Assessment (EIA) involves a thorough examination of particular environmental components that development projects might affect. A thorough assessment is essential to understand potential ecological impacts and facilitate informed decision-making. This chapter delves into the methodologies employed in these assessments, focusing on air quality, water resources, soil health, and biodiversity.

The chapter begins with an analysis of air quality, detailing common pollutants and their sources, and explores various technologies and methods for monitoring and assessing air quality. It then moves on to evaluate water resources, discussing techniques such as hydrological modelling and groundwater monitoring to determine water availability and quality. The examination of soil health includes evaluating properties like texture and nutrient content and identifying the impact of pollution and erosion. Finally, the chapter addresses biodiversity impacts, emphasising the importance of preserving natural ecosystems and exploring strategies for conservation. Throughout the chapter, legal standards and public participation are highlighted as integral to ensuring comprehensive and effective environmental assessments.

Air Quality Assessment

Assessing air quality is a critical component of the Environmental Impact Assessment (EIA) process, offering insights into potential health and ecological risks posed by development projects. This subpoint explores the methodologies employed and underscores their importance.

To begin, identifying major pollutants is fundamental. Common pollutants include particulate matter (PM), nitrogen oxides (NO_x), sulphur dioxide (SO_2), carbon monoxide (CO), volatile organic compounds (VOCs), and ground-level ozone (O_3). These pollutants primarily originate from industrial activities, construction, transportation, and energy production. For instance, construction sites often release PM due to soil disturbance and diesel exhaust from machinery. Similarly, factories may emit NO_x and SO_2 through combustion processes.

Understanding these pollutants' origins helps tailor specific mitigation strategies. For example, in urban development, implementing dust control measures can reduce PM emissions, while using cleaner fuels or energy-efficient equipment in factories can lower NO_x and SO_2 levels.

Technologies and methods for assessing air quality are diverse and have evolved significantly. One primary method involves ambient air monitoring, which collects and analyses air samples to determine pollutant concentrations. Devices such as high-volume samplers, gas analyzers, and remote sensing technology are integral to this process. High-volume samplers gather large air volumes over time, enabling detailed analysis of particulate concentrations. Gas analyzers measure specific gaseous pollutants, providing real-time data on air quality levels.

Another critical methodology is dispersion modelling. This technique uses mathematical models to simulate how pollutants disperse in the atmosphere from their sources. It accounts for factors like wind speed, topography, and atmospheric conditions, predicting pollutant concentrations at various locations. For this purpose, the U.S. Environmental Protection Agency (EPA) supports the AERMOD model. Dispersion modelling is invaluable for impact prediction and planning mitigation measures.

Additionally, mobile monitoring units offer versatility, allowing on-the-go data collection across different sites. These units are equipped with advanced sensors capable of detecting multiple pollutants, making them useful for comprehensive assessments in large project areas. Advancements in satellite technology also enable broader air quality monitoring, providing valuable data on regional and global scales.

Poor air quality has significant implications for human health and biodiversity. Exposure to high levels of PM can lead to respiratory issues, cardiovascular diseases, and premature death. NOx and SO2 can cause respiratory problems and aggravate pre-existing conditions like asthma. Ground-level O3, a secondary pollutant formed from reactions between VOCs and NOx under sunlight, poses severe respiratory hazards and can reduce lung function.

Biodiversity is equally affected. Pollutants like SO2 and NOx can lead to acid rain, which harms aquatic ecosystems by lowering water pH levels, disrupting species balance, and damaging vegetation. Ozone exposure can impair plant growth, affecting entire food webs and ecosystems. Sensitive species may decline, leading to reduced biodiversity and altered ecosystem functions.

Legal standards guide air quality assessments, ensuring uniformity and comprehensiveness. In many countries, air quality regulations are established based on guidelines from organisations like the World Health Organisation (WHO) and the EPA. These standards define permissible limits for various pollutants, ensuring that air quality levels remain within safe margins.

For example, the WHO Air Quality Guidelines set forth recommended limits for PM2.5, PM10, O3, NO2, SO2, and CO. These guidelines are based on extensive scientific evidence linking pollutant exposure to health outcomes. Regulatory bodies adopt these guidelines, tailoring them to local contexts and enforcing compliance through legislation.

The Clean Air Act in the United States is a robust framework regulating air emissions from stationary and mobile sources. It mandates the establishment of National Ambient Air Quality Standards (NAAQS) for pollutants considered harmful to public health and the environment. Compliance with NAAQS is monitored through state implementation plans, which require industries to adopt emission reduction technologies and practices.

In Europe, the Directive 2008/50/EC on ambient air quality sets similar standards for member states, emphasising the need for continuous monitoring and reporting. Non-compliance triggers corrective actions, including stricter emission controls and penalties.

Adhering to legal standards necessitates thorough and accurate assessments during the EIA. Baseline air quality data are essential for establishing a reference point against which project impacts are measured. Continuous monitoring throughout the project's life cycle ensures that air quality remains within acceptable limits, facilitating timely intervention if deviations occur.

Mitigation measures are devised based on assessment results, incorporating best practices and technological innovations. These measures aim to minimise pollutant emissions, improve air quality, and protect human health and biodiversity. For instance, employing dust suppression techniques, using low-emission construction machinery, installing air pollution control devices, and adopting cleaner production processes are effective strategies to mitigate negative impacts.

Public participation and transparency are crucial elements of air quality assessments in EIA. Involving stakeholders, including local communities, ensures that their concerns are addressed and they are informed about potential risks and mitigation efforts. Public consultations and access to assessment reports foster trust and compliance, enhancing the overall effectiveness of environmental management plans.

Water Resources Evaluation

Sustainable water management is a crucial aspect of project planning that ensures the longevity and health of water resources. Evaluating water availability and quality is the first step in this process. Techniques for assessing these factors include hydrological modelling, which predicts water flow and distribution within a watershed. Groundwater monitoring involves measuring water levels in wells over time and providing data on aquifer health and recharge rates. Additionally, water sampling for chemical analysis helps detect contaminants, offering insights into the water's suitability for various uses.

National and international standards for water quality assessment in Environmental Impact Assessments (EIA) provide a framework for consistent evaluation. Standards such as those established by the World Health Organisation (WHO) or the U.S. Environmental Protection Agency (EPA) define acceptable limits for pollutants like nitrates, heavy metals, and microbial contaminants. Adhering to these standards ensures that assessments are comprehensive and comparable across different regions and projects.

Over-extraction and pollution have significant impacts on aquatic life. Excessive withdrawal of water from rivers and lakes can lower water levels, disrupt habitats and reduce the availability of resources for species that depend on them.

Pollution from industrial discharge, agricultural runoff, and wastewater can introduce harmful substances into aquatic ecosystems, leading to issues like eutrophication, which depletes oxygen levels and harms fish and other marine life. The health of freshwater ecosystems is directly linked to the balance of water extraction and the prevention of pollution.

Involving local stakeholders in water resource assessments is essential for sustainable water management. Local communities often possess valuable knowledge about regional water sources, seasonal variations, and historical changes in water quality and availability. Engaging these stakeholders in the assessment process not only enriches the data collected but also fosters a sense of ownership and responsibility towards maintaining water resources. Public consultations, participatory mapping, and community-based monitoring programmes are effective methods for involving local voices in decision-making.

Hydrological assessments play a critical role in understanding water dynamics within a landscape.

This involves studying the movement, distribution, and quality of water in all its forms: surface water, groundwater, and atmospheric water. Through techniques such as Geographic Information System (GIS) mapping and remote sensing, hydrologists can create detailed maps and models that predict water behaviour under various scenarios. These assessments are vital for determining the sustainability of water use in project planning.

The importance of adhering to national and international water quality standards cannot be overstated. These standards provide benchmarks for evaluating the impact of proposed projects on water resources. For example, the European Union's Water Framework Directive sets a legal framework to protect and enhance the status of aquatic ecosystems and promote sustainable water usage. Compliance with these standards ensures that water quality assessments are rigorous and meet globally recognised criteria, promoting health and environmental protection.

Aquatic life suffers immensely from the consequences of over-extraction and pollution. Fish populations decline as reduced water levels affect breeding grounds and food availability. Pollutants like heavy metals and pesticides accumulate in the tissues of aquatic organisms, leading to bioaccumulation and biomagnification, where the concentration of toxins increases up the food chain. Addressing these impacts requires stringent control measures and regular monitoring to ensure that water bodies remain viable habitats for aquatic species.

Stakeholder engagement is critical in the context of water resource management. By involving local stakeholders, including indigenous groups and local governments, project planners can gain a comprehensive understanding of local water issues. Indigenous communities' traditional ecological knowledge frequently provides insights into sustainable practices that modern science might overlook. Workshops, public forums, and joint committees are practical ways to facilitate stakeholder participation, ensuring that their concerns and suggestions are integrated into the planning process.

Conducting thorough hydrological assessments as part of EIA illuminates the intricate connections between water resources and the environment. These assessments help identify potential risks associated with water shortages and flooding, guiding the implementation of mitigation measures. Techniques such as isotope hydrology, which uses the natural isotopes present in water molecules to trace water sources and movements, offer advanced tools for understanding the complexities of water cycles and their interactions with human activities.

Adhering to established water quality standards during EIAs ensures that the impact of development activities on water resources is assessed with precision. These standards serve as a reference point for evaluating the extent of contamination and identifying areas needing remediation. Global frameworks like the International Organisation for Standardisation (ISO) 14046, which provides guidelines for water footprint assessment, emphasise the importance of measuring and managing water use impacts throughout a project's lifecycle.

The detrimental effects of over-extraction and pollution on freshwater ecosystems underscore the need for robust regulatory measures and continuous monitoring. Legal instruments like the Clean Water Act in the United States mandate the restoration and maintenance of the integrity of the nation's waters by preventing point-source pollution and protecting wetlands. Effective enforcement of such regulations is crucial for preserving aquatic biodiversity and maintaining ecological balance.

Incorporating local stakeholders into water resource management processes fosters transparency and trust. Stakeholders' involvement ensures that the project planning aligns with the community's needs and values, reducing conflicts and facilitating smoother implementation. Collaborative approaches, where stakeholders actively participate in data collection, analysis, and decision-making, not only enhance the quality of assessments but also empower communities to take an active role in safeguarding their water resources.

Soil Health Analysis

Analysing soil health is critical in assessing the environmental impacts of development projects. Soil health directly affects plant growth, water filtration, and overall ecosystem stability. Understanding and maintaining healthy soil ensures that ecosystems remain functional and sustainable.

To begin with, several crucial properties of soil need to be evaluated, including texture and nutrient content. The texture of soil—whether it is sandy, clayey, or loamy—determines its drainage capabilities and root penetration possibilities. Sandy soils, for example, drain quickly but often lack nutrients, while clay soils retain water but may impede root growth. Loamy soils are generally considered ideal as they balance drainage and nutrient availability. Nutrient content is equally essential in assessing soil health. Key nutrients such as nitrogen (N), phosphorus (P), and potassium (K) play significant roles in plant growth and must be present in adequate quantities. Additionally, organic matter in the soil contributes to nutrient cycling and improves soil structure.

Assessing these properties involves various techniques, broadly categorised into chemical, physical, and biological tests. Chemical testing focuses on identifying the concentration of essential nutrients, pH levels, and potential contaminants like heavy metals. These tests help determine whether the soil can support plant life or if it requires amendments. Physical tests, on the other hand, include assessments of soil texture, structure, and compaction. Techniques such as the soil ribbon test, which involves rolling soil between fingers to gauge texture, and penetrometers, which measure soil compaction, provide valuable insights into the soil's physical condition. Biological tests are equally vital, focusing on the living organisms within the soil. Microbial biomass, earthworms, and root growth analyses offer indicators of a vibrant, functioning soil ecosystem.

Development activities pose significant threats to soil health, with pollution and erosion being primary concerns. Pollution from construction sites, industrial activities, and agricultural runoff can introduce harmful substances like heavy metals, pesticides, and excess nutrients into the soil. These pollutants can disrupt microbial communities, inhibit plant growth, and cause long-term damage to the ecosystem. Erosion, typically caused by deforestation, overgrazing, and construction, leads to the loss of topsoil, which contains most of the soil's organic matter and nutrients. Without this protective layer, the soil becomes less fertile and more prone to further degradation.

Effective strategies are essential to counteract soil degradation identified during assessments. To mitigate pollution, implementing best management practices (BMPs) is crucial. These include buffer strips of vegetation around construction sites to absorb runoff, proper disposal of hazardous materials, and using environmentally friendly agricultural practices such as integrated pest management (IPM). For erosion control, strategies like contour ploughing, terracing, and planting cover crops help stabilise the soil and prevent further loss. Contour ploughing involves tilling the soil along the natural contours of the land, reducing runoff and soil erosion. Terracing transforms steep slopes into stepped levels, making them more suitable for cultivation and reducing erosion rates. Cover cropping involves planting crops in between periods of regular crop production; these crops help protect the soil from erosion, improve nutrient content, and enhance soil structure.

Additionally, there are advanced technologies and approaches for monitoring and improving soil health. Remote sensing and geographic information systems (GIS) allow for large-scale monitoring of soil conditions, providing data on soil moisture, temperature, and vegetation cover. This technology helps identify areas at risk of degradation and prioritise them for conservation efforts. Soil amendments, such as adding organic compost or biochar, can also improve soil structure, increase nutrient availability, and boost microbial activity. Biochar, a form of charcoal produced from organic waste, has been shown to enhance soil fertility, retain water, and sequester carbon, contributing to both soil health and climate change mitigation.

Community involvement and education are vital components of successful soil health initiatives. Engaging local communities in soil conservation efforts not only raises awareness about the importance of soil health but also empowers individuals to take action in their own environments. Educational programmes can teach farmers and landowners sustainable land management practices, fostering a sense of stewardship and responsibility towards the land. Collaboration between government agencies, non-governmental organisations, and community groups can lead to more effective implementation of soil conservation strategies and policies.

Flora and Fauna Impacts

Assessing biodiversity is a crucial aspect of Environmental Impact Assessments (EIA), given the importance of preserving our natural ecosystems and understanding the profound impacts that developmental projects may have on them. This section delves into the methodologies and importance of assessing impacts on biodiversity, providing insights for environmental professionals, policymakers, and students alike.

Biodiversity Assessment Techniques

To effectively evaluate the flora and fauna affected by projects, several standardised methods are employed in biodiversity assessments. One such method involves field surveys, which include both direct observation and indirect evidence collection. Direct observation allows biologists to record existing species through sight or sound. This can be augmented with tools like camera traps, which capture images of elusive wildlife. In contrast, indirect evidence collection involves looking for signs of animal presence, such as tracks, nests, or droppings, contributing to a fuller understanding of the ecosystem.

Another widely used technique is the use of Geographic Information Systems (GIS) and Remote Sensing. These technologies map and analyse spatial data to identify changes in land use and habitat conditions over time. By integrating satellite imagery with ground-truthing—verifying satellite data with on-the-ground observations—researchers can detect patterns and trends that might impact biodiversity.

Molecular techniques, such as DNA barcoding, are gaining prominence in assessing biodiversity. DNA barcoding involves extracting genetic material from organisms and comparing it to known sequences in databases to accurately identify species. This method is particularly useful in identifying microorganisms and cryptic species that are not easily distinguishable through traditional means.

Impacts of Habitat Alteration

The development of projects, whether urban, industrial, or infrastructural, often leads to significant alterations in wildlife habitats. This fragmentation poses a severe threat to biodiversity. For instance, construction activities can disrupt natural corridors that animals use for migration, feeding, and breeding. When these pathways are blocked or narrowed, it can lead to isolated populations that are more vulnerable to inbreeding, disease, and extinction.

Moreover, habitat alteration can result in the loss of critical resources. Vegetation clearance for agriculture or housing developments reduces food availability and shelter for countless species. Aquatic habitats are equally impacted; water bodies may become polluted or altered by dam constructions, affecting fish populations and aquatic plants. Such disturbances ultimately reduce biodiversity resilience, making ecosystems less capable of withstanding changes or recovering from disturbances.

Conservation Strategies

To mitigate the negative impacts identified during EIA, conservation measures must be strategically implemented. One key strategy is the establishment of protected areas. By designating regions where human activity is restricted or managed, we can preserve crucial habitats and biodiversity hotspots. Buffer zones around these areas further help in minimising external pressures from nearby developments.

Restoration ecology offers another set of strategies. This involves rehabilitating degraded environments through reforestation, wetland restoration, and erosion control measures. Reforestation projects, for example, not only restore habitats but also sequester carbon, contributing to climate change mitigation efforts. Wetland restoration helps in maintaining water quality and providing habitat for various aquatic species.

Promoting sustainable development practices is vital. Environmental professionals can advocate for the incorporation of green infrastructure within project designs, such as creating wildlife corridors to maintain connectivity between habitats or utilising eco-friendly building materials to minimise the ecological footprint. Additionally, implementing adaptive management plans that allow for continuous monitoring and adjustment based on environmental feedback is crucial in ensuring long-term sustainability.

Ecological Value of Biodiversity

Maintaining healthy ecosystems holds intrinsic ecological and economic value. Ecologically, high biodiversity enhances ecosystem services, which are the benefits that humans derive from nature. These include pollination of crops, purification of air and water, and regulation of climate. Diverse ecosystems are also more resilient to environmental changes and disturbances, such as storms or diseases, because they contain a variety of species that can adapt and maintain ecosystem functions.

From an economic perspective, biodiversity contributes significantly to industries like agriculture, pharmaceuticals, and tourism. Many crops depend on pollinators like bees, whose decline could drastically affect food production and prices. The pharmaceutical industry relies on genetic diversity to discover new medicines. For instance, many life-saving drugs have been derived from compounds found in plant and animal species. Moreover, ecotourism generates substantial revenue for many regions, promoting conservation efforts while supporting local economies.

Integrated Impact Mitigation Strategies

Developing comprehensive strategies to mitigate various environmental impacts identified during Environmental Impact Assessments (EIA) is essential for sustainable development. This involves a multi-faceted approach addressing land, biodiversity, air, and water quality.

Promotion of sustainable land management practices begins with a thorough soil health analysis. Healthy soils are foundational to ecosystems as they support agriculture, prevent erosion, and filter water. To implement these practices, understanding soil characteristics such as nutrient content, texture, and pH levels is crucial. For instance, crop rotation and cover cropping can improve nutrient cycling and reduce soil erosion. Similarly, conservation tillage minimises disturbance to the soil structure, enhancing its organic matter content. The results from soil health analyses guide these practices, ensuring that management techniques align with the specific needs of the land. For example, areas identified with low organic matter may benefit from the addition of compost or manure to boost fertility and microbial activity.

Moving on, integrating conservation into project planning stages significantly aids in protecting biodiversity. Early inclusion of conservation measures ensures that potential impacts on ecosystems are considered before any ground is broken. For example, setting aside protected zones within a development area helps preserve habitats for local flora and fauna. Additionally, implementing green infrastructures like wildlife corridors and buffer zones facilitates safe movement for species across fragmented habitats. Restoration projects focused on re-establishing native vegetation can also counteract the ecological footprint of development activities. Moreover, environmental assessments should consider seasonal patterns of wildlife, breeding seasons, and migration routes to minimise disruptions.

Another cornerstone of an effective EIA strategy is the adoption of advanced monitoring equipment for precise air and water quality assessments. Modern technologies have dramatically improved our ability to measure environmental parameters accurately and in real-time. For air quality, sensors and satellite-based monitoring systems provide data on pollutants such as particulate matter (PM), nitrogen oxides (NOx), and sulphur dioxide (SO2). These tools enable continuous monitoring, capturing fluctuations and helping to pinpoint pollution sources quickly. On the water quality front, technological advancements like remote sensing and automated sampling stations offer detailed insights into parameters such as pH, temperature, dissolved oxygen levels, and contaminants. These sophisticated tools facilitate more immediate responses to negative trends, ensuring swift mitigation actions are taken.

In addition to technological solutions, involving local stakeholders in holistic water evaluations promotes sustainable water management. Participatory assessments engage community members, who often hold valuable traditional knowledge and have a vested interest in preserving their local environment. Stakeholder participation fosters transparency and builds trust between project developers and the community. Such involvement can include regular public meetings, workshops, and collaborative decision-making processes where residents contribute to water resource assessments. This inclusive approach helps ensure that water management strategies reflect both scientific findings and local realities.

For instance, local insights can identify critical water sources and cultural significance that might otherwise be overlooked. Engaging stakeholders early and throughout the lifecycle of a project enhances the long-term sustainability of water resources.

When implementing these comprehensive strategies, it's important to consider local regulations and guidelines. Sustainable land management practices should align with national agricultural guidelines and soil conservation policies. Conservation efforts must adhere to wildlife protection laws and international treaties, such as the Convention on Biological Diversity. Advanced monitoring technology should comply with environmental monitoring standards, ensuring data accuracy and reliability. Lastly, participatory approaches ought to follow frameworks for community engagement established by local governance bodies and international best practices.

Concluding Thoughts

This chapter has provided a comprehensive overview of the methods and significance of assessing specific environmental elements affected by development projects. By focusing on air quality, water resources, soil health, and biodiversity, we have highlighted how thorough evaluations can prevent or mitigate adverse ecological impacts. The methodologies discussed, such as air dispersion modelling, groundwater monitoring, soil texture analysis, and biodiversity field surveys, equip professionals with the necessary tools to make informed decisions. The integration of advanced technologies and regulatory frameworks ensures that assessments are both precise and effective, aligning with global standards for environmental protection.

Moreover, this chapter underscores the importance of stakeholder involvement and adherence to legal guidelines in conducting successful Environmental Impact Assessments (EIAs). Engaging local communities and incorporating traditional knowledge enriches the assessment process, fostering transparency and compliance. Legal standards like the WHO guidelines and national acts provide a uniform basis for evaluating and managing environmental impacts. In conclusion, the detailed exploration of these critical areas emphasises their interconnectivity and the need for holistic approaches to sustainable development, aiming at safeguarding both human health and ecological integrity.

Chapter 7

Social and Economic Considerations

Evaluating the social and economic impacts of development projects is essential for understanding their broader implications on communities. This chapter delves into various aspects, showcasing how thorough assessments can contribute to more informed decision-making processes. By examining both social and economic dimensions, the chapter aims to provide a comprehensive overview of the multifaceted effects that development initiatives can have on resource management and community well-being.

The chapter addresses several key topics to illuminate the importance of integrating social and economic considerations into development planning. First, it explores methodologies for assessing social impacts, highlighting tools and techniques used to gauge changes in community cohesion, public health, and overall quality of life. Next, it examines economic evaluations, focusing on how projects influence local economies, employment rates, and income levels. Additionally, the chapter covers the role of stakeholder engagement, emphasising the significance of including community voices in the assessment process. Finally, it presents case studies to illustrate successful strategies and lessons learned from past projects, providing practical insights for practitioners and policymakers.

Health Impact Assessments

Assessing the health impacts of development projects is crucial to ensuring community well-being. Health impact assessments (HIAs) perform systematic evaluations of potential health risks associated with such projects. By conducting thorough HIAs, developers can identify a range of health issues that might arise due to construction activities, changes in environmental conditions, or increased population density. These assessments help to spot health-related problems before they become significant, thereby allowing for appropriate adjustments and mitigating measures within project plans.

Health impact assessments are comprehensive tools designed to analyse various factors that could influence public health. For example, they consider air quality, water quality, noise levels, and socio-economic changes that accompany development projects. HIAs scrutinise how pollutants from construction may affect respiratory health or how changes in traffic patterns might increase accident rates. This multidimensional analysis provides a detailed picture of potential health risks, enabling stakeholders to make informed decisions.

Conducting clear guidelines for health assessments significantly improves public health outcomes. Guidelines ensure that all relevant health aspects are thoroughly examined and that standardised methods are employed for data collection and analysis. These guidelines usually encompass steps such as screening, scoping, assessing risks, recommending actions, and monitoring outcomes. They also specify the types of data to be collected, which methodologies to employ, and how to interpret the findings. By adhering to these established procedures, HIAs can deliver reliable and consistent results, which form the basis for making effective health-related recommendations.

Timely identification of possible health risks allows for early interventions during the project planning stages. When health risks are identified early, developers have the opportunity to incorporate preventive measures into their design and operational plans. For example, if an HIA reveals that a new road might deteriorate local air quality to harmful levels, project planners can modify routes, propose green belts, or introduce pollution control technologies. Early identification and timely response reduce adverse health impacts and result in better public health protection.

Effective health impact assessments also foster trust between developers and communities.

Transparency and thoroughness in the HIA process demonstrate a commitment to community well-being, encouraging public engagement and support. When communities see that developers prioritise their health concerns and implement safeguards accordingly, it builds confidence in the project and reduces opposition. Public meetings, workshops, and openly shared HIA results allow residents to voice their concerns, contribute local knowledge, and feel involved in decision-making processes, thus creating a collaborative environment.

Methods for conducting HIAs vary but generally follow a structured approach. The initial step involves screening, which helps to determine whether an HIA is necessary for the given project. This step sets the stage for more detailed analysis, focussing resources on projects that potentially pose significant health risks. Following screening, the scoping phase identifies what health effects will be considered and who will be affected. This phase is critical for defining the HIA's scope and ensuring no key health issues are overlooked.

Assessment and risk evaluation constitute the heart of the HIA process. During this phase, data from multiple sources, such as scientific literature, health statistics, and community input, is collected and analysed. Quantitative and qualitative methods often complement each other, providing a robust evaluation of potential health impacts. This comprehensive assessment allows for a nuanced understanding of how different elements of the project might affect public health.

Recommendations derived from HIAs are actionable measures aimed at mitigating identified health risks. These might include modifying project designs, implementing health promotion programmes, or enhancing environmental management practices. Recommendations should be practical, evidence-based, and tailored to the specific context of the project and community. Effective communication of these recommendations to stakeholders ensures they are incorporated into project plans and implementation strategies.

Monitoring and evaluation are integral to the HIA framework, ensuring that recommended measures are effective and health outcomes are as predicted. Continuous monitoring allows for adjustments to be made in response to unforeseen health issues that may arise during project execution. It also provides valuable feedback for improving future HIAs, contributing to an evolving body of knowledge and practice in health impact assessment.

The inclusion of community perspectives in the HIA process enhances its relevance and effectiveness. Engaging with community members through surveys, focus groups, and public consultations brings in local insights and values that might not be evident through purely technical analyses. This participatory approach ensures that the project aligns with community health priorities and addresses specific concerns, fostering a sense of ownership and cooperation among residents.

Real-world examples illustrate the tangible benefits of integrating HIAs into project planning. For instance, an HIA conducted for a major highway construction project might reveal potential increases in asthma cases among children living nearby due to elevated air pollution levels. Armed with this knowledge, planners could introduce measures such as enhanced ventilation systems in nearby schools, planting trees to act as air filters, or rerouting heavy traffic away from residential areas. Such proactive steps demonstrate the utility of HIAs in safeguarding public health.

Safety and Risk Evaluations

The evaluation of safety and risk is a fundamental aspect that can significantly influence both community acceptance and the overall viability of development projects. Understanding these elements begins with defining what constitutes safety and risk evaluations, as they are essential for assessing potential hazards associated with any development endeavour.

Safety evaluations involve examining the potential dangers and ensuring adequate measures are in place to protect both individuals and properties from harm. This involves identifying possible sources of harm, such as structural failures, environmental accidents, or operational mishaps, and developing strategies to mitigate these risks. Risk evaluations, on the other hand, involve assessing the likelihood and potential severity of these hazards occurring. By evaluating risks, project planners can prioritise which hazards need immediate attention and resource allocation.

One crucial technique in this domain is quantitative risk analysis. Utilising quantitative methods allows for a more precise and objective assessment of risks. These methods typically involve mathematical models that estimate the probability of different risks and their potential impacts. For example, a quantitative risk analysis might use statistical data to estimate the likelihood of equipment failure or the probability of natural disasters affecting the project site. This concrete data enables decision-makers to make informed choices about design modifications, resource allocation, and emergency preparedness plans.

Quantitative analysis also facilitates cost-benefit calculations, helping to determine whether the benefits of certain safety measures justify their costs.

Additionally, community input plays an indispensable role in tailoring safety measures to meet specific local needs. Community members often possess valuable local knowledge that can identify unique risks not immediately apparent to external experts. Engaging with communities through public consultations, surveys, and town hall meetings provides insights into community concerns and expectations. This collaborative approach not only helps in designing more effective safety measures but also builds trust between developers and residents. When communities feel heard and see their concerns addressed, they are more likely to support development projects.

Continuous risk assessments throughout the life cycle of a development project are critical to ensuring emerging risks are managed effectively. Initial risk assessments provide a snapshot of potential hazards at the planning stage, but risks can evolve as projects progress. New technologies, changing environmental conditions, and operational changes can introduce new risks or alter the nature of existing ones. Therefore, ongoing risk assessments are necessary to monitor these developments and adjust safety measures accordingly. Regularly scheduled assessments, along with real-time monitoring systems, allow for the early detection of new risks and the implementation of timely interventions.

A practical guideline for conducting comprehensive safety and risk evaluations includes several key steps. First, establish a robust risk management framework that outlines the processes for identifying, assessing, and mitigating risks. This framework should be adaptable to various types of projects and scaleable based on project size and complexity. Second, employ a combination of qualitative and quantitative risk assessment methods to gather a complete picture of potential hazards. While quantitative methods provide concrete data, qualitative methods such as expert interviews and scenario analyses offer context and deeper understanding. Third, engage with all relevant stakeholders, including local communities, regulatory bodies, and industry experts, to ensure diverse perspectives are considered. Fourth, implement risk mitigation strategies based on the assessment findings, prioritising actions that address the most significant risks first. Finally, establish continuous monitoring protocols to track the effectiveness of implemented measures and identify any emerging risks promptly.

For example, in a construction project near a seismic fault line, safety evaluations would initially focus on designing structures to withstand earthquakes. Quantitative risk analysis might involve calculating the likelihood of various earthquake magnitudes and their potential impact on the structures. Community input could reveal concerns about evacuation routes and emergency shelters. Continuous risk assessments would then monitor geological activity and update building codes and emergency plans as necessary.

Analysis of Livelihoods

Evaluating the potential impacts of development projects on community livelihoods is essential for ensuring that local economies are considered in impact assessments. By doing so, we can offer a balanced perspective that considers both economic growth and social well-being. Introducing frameworks for analysing these impacts provides the necessary structure for a thorough and nuanced approach.

Frameworks such as the Sustainable Livelihood Framework (SLF) offer a comprehensive method to examine the impacts on community livelihoods. SLF focuses on how people use a combination of assets—natural, human, financial, social, and physical capitals—to achieve their livelihood goals. Incorporating this framework allows practitioners to assess multiple dimensions of livelihoods and understand how various factors interact. This holistic view ensures that all aspects of community life are taken into account, leading to more balanced and effective decision-making.

A key aspect of evaluating project impacts is the combination of quantitative surveys and qualitative interviews. Quantitative surveys provide numerical data that can highlight trends and project impacts on income, employment rates, and access to resources. For instance, detailed economic surveys can measure changes in household income and employment levels before and after a project's implementation. These metrics are crucial for identifying whether a project has positive or negative economic impacts on a community.

On the other hand, qualitative interviews offer insights into people's experiences, perceptions, and concerns. They allow us to capture stories and voices that quantitative data might overlook. For example, interviewing local business owners can reveal how new developments affect customer foot traffic, supply chains, and overall business viability. Combining these methods provides a multi-dimensional view, offering both the breadth of quantitative data and the depth of qualitative insights. This mixed-method approach enables more informed and empathetic evaluations.

Learning from past projects plays a critical role in understanding common pitfalls and successes. Historical analysis allows us to identify patterns and outcomes that can inform current practices. For example, examining case studies of previous infrastructure projects can highlight recurring issues such as displacement, loss of traditional livelihoods, or unintended economic disparities. Identifying these patterns helps practitioners design better mitigation strategies and avoid repeating mistakes.

In addition, past success stories offer valuable lessons. Analysing successful community engagement initiatives illustrates how inclusive planning can lead to more sustainable outcomes. For instance, examining a successful renewable energy project might reveal how involving local stakeholders in the decision-making process fostered community support and ensured project sustainability. By learning from both failures and successes, we build a repository of knowledge that guides future efforts.

Proactive planning is another essential component for helping businesses adapt and thrive amidst project changes. Anticipating potential challenges and developing strategies in advance can mitigate negative impacts. For example, when planning large construction projects, it's crucial to identify how local businesses might be affected by disruptions like road closures or changes in customer access. Proactively communicating with business owners and providing them with contingency plans can help minimise these impacts.

Furthermore, supporting local businesses during transitional phases is vital. Offering training programmes, financial assistance, or business development services can equip them to navigate changes effectively. For instance, if a new industrial project leads to an influx of workers, local businesses might need help scaling up their operations to meet increased demand. Providing targeted support enables businesses to seize new opportunities and contribute positively to the local economy.

Guidelines for effective data collection methods ensure rigorous and reliable assessments. For quantitative surveys, establish clear objectives and methodologies. Use representative sampling techniques to ensure that data accurately reflects the broader community. Additionally, ensure that surveys are designed to capture relevant economic indicators such as income levels, employment status, and access to resources.

For qualitative interviews, develop structured interview guides to maintain consistency while allowing flexibility for respondents to share their experiences. Train interviewers to build rapport and ask probing questions that uncover deeper insights. Employ various data validation methods, including peer reviews and triangulation, to enhance the credibility of findings.

Implementing these guidelines helps produce high-quality data, forming the basis for sound decision-making. Reliable data enables a clearer understanding of project impacts, informing the development of strategies that address identified issues effectively.

Community Cohesion Impacts

Development projects often bring significant changes to communities, influencing their social structures and cohesion. Understanding how these projects impact community cohesion is crucial for addressing potential conflicts that may arise. When development projects are planned or initiated without considering existing social frameworks, disruptions can occur, leading to disputes among community members and between the community and developers. By thoroughly assessing these impacts, project planners can take proactive steps to ensure harmonious interactions within the community.

Community cohesion refers to the strength of relationships and the sense of solidarity among members of a community. It encompasses trust, mutual respect, shared values, and cooperation. Assessing community cohesion helps identify areas where development might create tension or discord. For instance, if a project displaces residents or alters communal spaces, it could weaken social bonds and increase conflicts. Conversely, if developments provide new amenities or improved infrastructure, they may strengthen community ties. Therefore, understanding these dynamics is fundamental in creating development projects that foster positive social outcomes.

One effective method to measure community cohesion is through surveys and social mapping techniques. Surveys can gather quantitative data on various aspects of community life, such as levels of trust, perceptions of safety, participation in community activities, and satisfaction with local services. These surveys should be carefully designed to capture diverse voices within the community, ensuring that all demographic groups are represented. Social mapping, on the other hand, visually represents relationships and networks within the community, highlighting key social hubs and potential areas of isolation. This technique allows developers to understand the geographical distribution of social ties and identify critical areas that need attention during project planning.

Documented outcomes from these assessments provide valuable insights into how development projects either foster or hinder community interactions. By analysing past projects, developers can learn what strategies promote cohesion and which actions lead to fragmentation. For example, a successful case might involve a project that incorporated community feedback early on, resulting in shared facilities that enhanced social interactions. Alternatively, a project that ignored social input might have led to the creation of barriers between different sections of the community, exacerbating divisions. Such documentation serves as a guide for future projects, illustrating best practices and common pitfalls.

Building social ties through community engagement initiatives is another vital strategy to mitigate resistance to development projects. Engaging the community means involving them in the decision-making process, seeking their input, and addressing their concerns. This participatory approach not only empowers residents but also fosters a sense of ownership and collaboration. For example, organising town hall meetings, focus groups, and workshops can help bridge the gap between developers and community members. These forums provide opportunities for open dialogue, where developers can explain the project's goals and benefits while listening to the community's needs and suggestions. Additionally, collaborative activities such as community-led planning committees or volunteer programmes can strengthen social bonds and reduce resistance to development projects.

Effective communication is the cornerstone of successful community engagement. Clear, transparent, and consistent communication helps build trust and ensures that community members feel informed and valued. Developers should use multiple channels to reach different segments of the community, including traditional media, social media, and face-to-face interactions. Providing regular updates on project progress and addressing concerns promptly can alleviate fears and uncertainties, fostering a more cooperative environment.

Furthermore, developers should consider the cultural context of the community when planning and implementing engagement initiatives. Recognising and respecting cultural norms, traditions, and values can enhance the effectiveness of these efforts. For instance, in communities with strong oral traditions, storytelling sessions might be a more appropriate and impactful way to share information and gather feedback than written surveys. Cultural sensitivity also means being aware of power dynamics within the community and ensuring that marginalised groups have a voice in the process.

Beyond initial engagement, ongoing support and investment in community-building activities are essential for maintaining cohesion throughout the project's lifecycle. This can include funding for local events, supporting community organisations, and creating spaces for social interaction. For example, developing public parks, community centres, or shared gardens can provide venues for residents to come together, reinforcing social ties. Additionally, partnerships with local businesses and institutions can create opportunities for employment and economic growth, further integrating the project into the fabric of the community.

Recommendations and Strategies

Incorporating health, safety, risk, livelihood, and community cohesion considerations into Environmental Impact Assessments (EIAs) is critical for ensuring that development projects are beneficial and sustainable. This section outlines several strategies to achieve this integration effectively.

One of the first steps in achieving comprehensive EIAs is establishing mandatory assessments within regulatory frameworks. Making these evaluations a requirement helps standardise practices across different regions and sectors. When these assessments are not optional, developers must consider a broad range of potential impacts, leading to more thorough and consistent evaluations. For instance, many countries have enacted laws that require environmental assessments before any major project can proceed. These regulations create a baseline for what must be evaluated, including health risks, safety issues, economic impacts, and social dynamics. Standardised regulatory requirements ensure that all projects undergo a rigorous review process, which reduces the likelihood of overlooking critical factors that can affect communities.

Training stakeholders in assessment methodologies is another essential strategy. Effective training programmes enhance the understanding and execution of impact assessments among various stakeholders, including government officials, developers, consultants, and community members. Well-trained personnel are better equipped to carry out detailed and accurate assessments, identify potential risks early on, and propose effective mitigation measures. Training also fosters a common language and set of expectations among diverse stakeholders, which can facilitate collaboration and improve the overall quality of the assessments. For example, capacity-building workshops and courses on specific methodologies like Geographic Information Systems (GIS) or Health Impact Assessments (HIA) can equip stakeholders with the technical skills necessary for conducting comprehensive evaluations.

Continuous monitoring of projects is crucial for maintaining their adaptability over time. Initial assessments provide valuable insights, but ongoing monitoring ensures that unforeseen impacts or changes in circumstances can be addressed promptly. By keeping track of various indicators throughout the project lifecycle, stakeholders can identify emerging issues and adapt their strategies accordingly. This approach allows for a dynamic response to potential problems, ensuring that projects remain viable and beneficial to communities. For example, post-implementation reviews and regular environmental audits can help detect shifts in local ecosystems, socioeconomic conditions, or public health trends, enabling timely interventions. Continuous monitoring also builds trust within communities, as they see a commitment to long-term sustainability and accountability from the developers.

Policies that support economic diversification and community-led initiatives are pivotal in enhancing sustainability. Development projects often bring about significant changes to local economies, affecting livelihoods and community structures. Supporting economic diversification means promoting a variety of income-generating activities to reduce dependency on a single industry. This approach makes communities more resilient to economic shocks and enhances overall stability. Additionally, encouraging community-led initiatives empowers local populations to take an active role in the development process, fostering a sense of ownership and ensuring that projects align with their needs and aspirations. For instance, policies might include funding for small business development, vocational training programmes, or initiatives that promote local entrepreneurship. Community-led projects, such as cooperative farming or local tourism ventures, can also be instrumental in creating sustainable, locally-driven economic growth.

Insights and Implications

This chapter has explored the varied social and economic impacts of development projects, emphasising how rigorous assessments inform decision-making that benefits communities and promotes sustainable resource management. By thoroughly understanding health impacts, safety risks, effects on livelihoods, and community cohesion, practitioners can identify potential challenges and opportunities early in the planning stages. These evaluations enable developers to make adjustments that mitigate negative consequences while bolstering positive outcomes, ensuring that projects align with both local needs and broader sustainability goals.

Integrating comprehensive assessments into project planning fosters trust and cooperation between developers and communities. Transparent evaluation processes and proactive engagement with residents allow for the incorporation of valuable local insights, enhancing the relevance and effectiveness of development initiatives. Continuous monitoring and adaptive strategies ensure that projects remain resilient and beneficial over time. Implementing these holistic approaches equips stakeholders with the tools needed to navigate complexities, ultimately leading to more equitable and sustainable development practices.

Chapter 8

Principles of Sustainable Development

Principles of sustainable development are essential for guiding projects towards maintaining a balance between economic growth, environmental health, and social well-being. The core idea is that development activities should not compromise the ability of future generations to meet their own needs. This chapter delves into how Environmental Impact Assessment (EIA) can be utilised as a tool to achieve sustainability in project planning and implementation. Environmental professionals, policymakers, and students will find methodologies and practical tools that emphasise the importance of sustainability principles throughout this assessment process.

The chapter explores various key principles vital to sustainable development. It begins by discussing the concept of integration, ensuring that different sectors and scales work together harmoniously. Inclusiveness is also covered, highlighting the necessity of involving all stakeholders, especially marginalised communities, in decision-making. Resilience is another focus, showing the need for adaptable systems that can withstand economic, social, and environmental shocks. These fundamentals are linked to the Sustainable Development Goals (SDGs), providing a framework that reinforces global efforts towards sustainability. Practical examples and detailed methodologies embedded in the EIA process are examined to illustrate these principles in action, equipping readers with the knowledge to apply these concepts in real-world scenarios.

The Concept of Sustainable Development

Sustainable development is a multifaceted concept that lies at the intersection of economic growth, environmental stewardship, and social equity. It seeks to create a balanced approach where each of these dimensions supports and enhances the others, rather than existing in isolation or conflict. The concept was prominently defined in the Brundtland Report in 1987 as "development that meets the needs of the present without compromising the ability of future generations to meet their own needs." This definition underscores the importance of mindful consumption and resource management to ensure long-term viability for both human society and the planet.

At its core, sustainable development promotes a holistic approach to resource management, emphasising that all elements of development—economic, social, and environmental—must be considered collectively. By integrating these components, sustainable development ensures that policies and practices do not favour short-term gains over long-term sustainability. This approach acknowledges that economic activities inevitably impact the environment and society, thus promoting a more integrated strategy that considers a wide array of factors and stakeholders.

One crucial aspect of sustainable development is its reliance on several key principles: integration, inclusiveness, and resilience. Integration refers to the need to consider the interconnections between different sectors and scales of action, ensuring that decisions made in one area do not produce negative consequences in another. For instance, in implementing an Environmental Impact Assessment (EIA), it is essential to evaluate the potential economic benefits of a project alongside its environmental impacts and social implications. This comprehensive assessment helps prevent harmful trade-offs and aligns various aspects of development.

Inclusiveness is another fundamental principle, emphasising the involvement of all stakeholders in the decision-making process. Sustainable development recognises that marginalised groups often bear the brunt of environmental degradation and economic disparities. Therefore, inclusive practices aim to ensure that these communities are heard and considered in the planning and implementation phases of projects. This inclusivity fosters greater equity and social cohesion, contributing to more robust and just outcomes.

Resilience is about building systems that can withstand and adapt to shocks and stresses, whether they are economic downturns, social upheavals, or environmental disasters. Resilient systems are designed to absorb impacts while maintaining functionality and facilitating recovery. For instance, in project planning, incorporating adaptable designs and technologies can help mitigate the adverse effects of climate change, ensuring that communities continue to thrive despite changing conditions.

A critical element linking the principles of sustainable development with practical outcomes is the framework of Sustainable Development Goals (SDGs). The SDGs, which the United Nations established in 2015, offer a global agenda for sustainable practices by setting targets across 17 goals to address issues like poverty, inequality, climate change, and environmental degradation. These goals serve as a common blueprint for countries and organisations worldwide, guiding efforts towards achieving sustainable development.

The SDGs encourage a multidimensional approach to sustainability that ties local project outcomes to broader environmental and societal impacts. For example, Goal 13 emphasises urgent action to combat climate change and its impacts, prompting projects to incorporate climate resilience into their designs and operations (United Nations, 2024). Similarly, Goal 15 focusses on protecting terrestrial ecosystems and advocating for sustainable land use and forest management to halt biodiversity loss. By aligning project objectives with these global targets, practitioners can contribute to a larger movement towards sustainability, amplifying their efforts' positive effects.

Moreover, the SDGs emphasise the principle of partnership, recognising that achieving sustainable development requires collaboration across sectors and borders. Goal 17 calls for strengthening the means of implementation and reviving global partnerships. This goal highlights the importance of cooperation among governments, private businesses, civil society, and international organisations. Such partnerships facilitate the sharing of knowledge, resources, and best practices, fostering innovation and amplifying impact.

In practice, enhancing the EIA process through the lens of sustainable development means embedding these principles and goals into every stage of project planning and implementation. This approach involves conducting thorough assessments that account for potential long-term impacts, engaging a wide range of stakeholders in meaningful dialogue, and ensuring that projects are designed to be adaptable and resilient to future challenges. Additionally, aligning project strategies with the SDGs ensures that local actions contribute to global sustainability efforts, creating a coherent and unified approach to development.

Several real-world applications serve as examples of the holistic approach that sustainable development promotes. For instance, urban planning projects that integrate green infrastructure not only provide economic benefits through job creation and tourism but also enhance environmental quality by reducing pollution and increasing biodiversity. Socially, these projects offer recreational spaces and improve public health outcomes, demonstrating the interconnected benefits of considering economic, environmental, and social dimensions together.

Another example is in renewable energy projects, where integrating solar or wind power installations addresses multiple sustainability principles. Economically, these projects reduce dependence on fossil fuels and stabilise energy costs. Environmentally, they decrease carbon emissions and mitigate climate change impacts. Socially, they can improve energy access for underserved communities, fostering greater equity and inclusion.

Precautionary Principle

The precautionary principle is a critical concept in guiding Environmental Impact Assessments (EIAs) to prevent environmental harm through proactive measures. This principle is particularly significant when there is uncertainty regarding potential harm, as it calls for action even in the absence of complete scientific proof and shifts the burden of proof to the proponents of potentially harmful activities.

Definition and Importance

The precautionary principle mandates that measures should be taken to avoid or mitigate potential hazards to human health or the environment when scientific evidence about an environmental or health risk is uncertain. Essentially, it promotes the idea of being "better safe than sorry" and emphasises preventive over reactive measures. It shifts the responsibility from the public to prove harm to the proponent of the activity to prove the absence of harm. By doing so, it aims to protect both the environment and public health preemptively rather than waiting for damage to occur.

One notable definition stems from the Wingspread Consensus Statement on the Precautionary Principle, which states: "When an activity raises threats of harm to human health or the environment, precautionary measures should be taken even if some cause-and-effect relationships are not fully established scientifically" (*Precautionary Principle: An Overview | ScienceDirect Topics*, 2013). This reflects the importance of proactive intervention.

Application in Environmental Assessments

In the context of EIAs, the precautionary principle plays a vital role in promoting thorough risk assessments and adaptive management measures. During an EIA, identifying potential risks and uncertainties is crucial. This principle encourages comprehensive evaluation methods that consider the worst-case scenarios and potential long-term impacts of proposed projects.

For instance, conducting a detailed risk assessment under this principle involves evaluating various alternatives, including the possibility of taking no action at all. These evaluations often use tools like Bayesian networks, which use uncertain causal reasoning to show what might happen if different choices are made (Precautionary Principle: An Overview | ScienceDirect Topics, 2013). Moreover, adaptive management measures ensure that as new information emerges, project plans can be adjusted to mitigate unforeseen risks effectively.

Case Studies Highlighting Effectiveness

Several case studies highlight the effectiveness of the precautionary principle in averting large-scale environmental impacts. One prominent example is the European Union's approach to regulating genetically modified organisms (GMOs). The EU has adopted stringent regulations based on the precautionary principle, leading to rigorous assessments before any GMO products can be approved for cultivation or sale. This proactive stance has helped prevent potential ecological disruptions and preserved biodiversity.

Another example is New Zealand's early response to the COVID-19 pandemic. Despite having little knowledge of the virus's spread and effects, the government imposed strict travel restrictions and mandatory lockdowns under the precautionary principle. These early interventions were credited with significantly reducing the number of infections and deaths compared to countries that delayed such measures (Pinto-Bazurco, 2020).

Challenges and Criticism

Despite its benefits, the precautionary principle faces challenges and critiques, especially concerning conflicts with economic interests and misinterpretations that lead to overly cautious approaches. Critics argue that the principle can hinder technological innovation and development by imposing excessive regulatory burdens on industries. They contend that the cost of precautionary measures may outweigh the benefits, particularly when the risks are speculative or minimal.

Furthermore, there is the issue of balancing precaution with economic growth. Implementing precautionary actions often requires significant financial investments, which may strain resources, particularly in lower-income regions. For example, developing countries might face difficulties applying the principle due to limited technological capabilities and financial constraints (*Precautionary Principle: An Overview | ScienceDirect Topics*, 2013).

Misinterpretations of the principle can also lead to overly cautious approaches that stall beneficial projects. For instance, some policymakers may reject innovative solutions outright due to perceived risks without thoroughly assessing the potential benefits. This can result in missed opportunities for advancements in sustainability and environmental protection.

Polluter Pays Principle

The polluter pays principle (PPP) is a cornerstone of modern environmental policy, defined by its core idea that those who cause pollution must bear the costs associated with managing it to prevent damage to human health or the environment. This principle is significant for ensuring accountability and incentivizing pollution reduction within Environmental Impact Assessments (EIAs).

Concept and Significance : The polluter pays principle ensures that the financial burden of pollution does not fall on society but remains with the polluter. By linking the costs directly to the activities that generate pollution, it motivates industries and individuals to adopt cleaner production methods. When polluters are held financially responsible, there is a direct economic incentive to reduce emissions, invest in cleaner technologies, and manage waste more efficiently. This alignment of economic responsibility with environmental accountability promotes more sustainable practices and reduces the externalisation of environmental costs.

Implementation in EIA Processes : Within EIAs, the integration of the polluter pays principle mandates comprehensive cost-benefit analyses that include environmental costs. This approach ensures transparency and accountability in identifying and addressing environmental liabilities. By requiring project proponents to account for potential environmental damages in their financial planning, EIAs foster more responsible project designs. This process often involves detailed assessments of potential impacts, mitigation strategies, and the long-term sustainability of proposed activities. Moreover, this transparency helps stakeholders understand the full extent of a project's environmental footprint, promoting informed decision-making that aligns with sustainability goals.

Global Examples : Various countries have adopted the polluter pays principle with notable success, demonstrating its effectiveness in changing industrial behaviours and promoting environmental justice. For instance, the Waste Directive of 2008 in the European Union enforces the principle by requiring that the waste holder or the product's producer bear the costs of waste disposal. This regulation has led to significant improvements in waste management practices across member states, encouraging industries to minimise waste generation and explore recycling options. Similarly, in South Africa, the implementation of the PPP within water management policies has resulted in industries adopting more efficient water usage practices and investing in technologies to reduce wastewater discharge, thereby benefiting both the environment and local communities.

Another example is Japan's experience with industrial pollution in the mid-20th century, leading to the enactment of strict environmental regulations based on the polluter pays principle. Companies were required to compensate affected communities and invest in pollution control measures. This shift not only improved environmental conditions but also spurred technological innovation in pollution control and resource management. These global examples highlight how the polluter pays principle can drive positive environmental outcomes and support broader efforts towards sustainable development.

Controversies and Limitations : Despite its advantages, the polluter pays principle faces several challenges and controversies. One major issue is resistance from certain economic sectors, particularly those heavily reliant on traditional industrial processes. These sectors often argue that the additional costs imposed by the principle can undermine their competitiveness and profitability, potentially leading to job losses and adverse economic impacts. This resistance is particularly pronounced in regions where regulatory frameworks are less stringent and enforcement mechanisms are weak. Another significant limitation is the issue of fairness. While the principle aims to hold polluters accountable, it can disproportionately affect low-income communities and small businesses. These groups may lack the financial resources to invest in cleaner technologies or mitigate their environmental impact, making them vulnerable to punitive measures.

Furthermore, implementing the polluter pays principle in developing countries presents unique challenges. In these regions, regulatory infrastructure may be lacking, making it difficult to enforce compliance effectively. Moreover, the identification of actual polluters can be complex, especially in cases involving diffuse pollution sources. For instance, agricultural runoff, which contributes significantly to water pollution, is challenging to attribute to specific individuals or entities. This complexity necessitates robust monitoring systems and collaborative approaches involving multiple stakeholders to ensure effective implementation. Additionally, the application of the principle in multinational contexts raises questions about equity and historical responsibility. Industrialised nations, having contributed significantly to global pollution over the decades, face calls to provide financial support to developing countries struggling with the consequences of climate change and environmental degradation.

Balancing these historical responsibilities with contemporary enforcement of the polluter pays principle requires nuanced policy frameworks that address both immediate and long-term environmental justice concerns.

Intergenerational Equity

Intergenerational equity is a foundational concept in the realm of sustainable development, advocating for fairness between current and future generations. This principle emphasises that today's actions should not compromise the ability of future generations to meet their own needs. By integrating intergenerational equity into project planning, we can ensure that long-term resource availability is maintained and the benefits and burdens of development are distributed equitably over time.

The implications of intergenerational equity for Environmental Impact Assessments (EIA) are profound. EIAs serve as tools to evaluate the environmental consequences of proposed projects before they proceed. By incorporating intergenerational equity principles, EIAs can guide sustainable development strategies that consider the long-term impacts on future generations. This includes ensuring that resources are used in a manner that preserves their availability for the future and that the benefits and burdens of development are shared fairly across generations. This approach helps to avoid scenarios where short-term gains lead to long-term environmental degradation, thereby promoting a more balanced and sustainable development trajectory.

Real-world examples highlight the practical application and benefits of intergenerational equity. One notable instance is the management practices of indigenous communities, which often emphasise the sustainable use of natural resources. These practices have shown considerable success in maintaining ecological balance and providing socio-economic benefits. For example, in New Zealand, the Maori's traditional land management systems focus on preserving the land for future generations, ensuring that natural resources remain abundant and accessible. These practices demonstrate how stakeholder engagement, particularly involving those who will inherit the outcomes, can lead to better environmental and social results.

Another example is the implementation of national policies that integrate sustainability goals. Countries like Sweden and Costa Rica have enacted laws that mandate long-term environmental planning and conservation efforts. In Sweden, the Environmental Code includes provisions that protect the interests of future generations, while Costa Rica has achieved significant progress in renewable energy adoption and forest conservation. These examples show that policy frameworks rooted in intergenerational equity not only enhance environmental preservation but also foster social and economic stability.

Despite these successes, there are substantial future challenges to achieving intergenerational equity in sustainable development. One major challenge is the complexity of forecasting future needs and preferences. Predicting the requirements of future generations is inherently uncertain, making it difficult to plan effectively. For instance, technological advancements and shifts in societal values can drastically change what future generations may need or prioritise. Therefore, flexibility and adaptability must be built into sustainable development strategies to accommodate these potential changes.

Immediate economic pressures also pose significant obstacles. There is often a tension between short-term economic gains and long-term sustainability goals. Projects that promise immediate economic benefits might be prioritised over those that offer long-term environmental and social advantages, potentially leading to resource depletion and environmental harm. Policymakers and planners must navigate these economic pressures carefully, balancing the need for immediate economic development with the imperative of preserving resources and opportunities for future generations.

Another challenge lies in the gaps within current policy frameworks. Many existing policies do not adequately address the needs of future generations or fail to incorporate intergenerational equity principles comprehensively. This gap can result in fragmented and short-sighted decision-making processes. Developing more inclusive and forward-thinking policies requires a concerted effort to integrate long-term perspectives into all levels of planning and governance. This might include revising existing regulations to explicitly consider the impacts on future generations and creating new institutional structures that promote intergenerational dialogue and participation in decision-making.

Strategies to overcome these challenges include institutionalising long-term planning processes. Governments and organisations can establish dedicated bodies or committees focused on intergenerational equity to ensure that long-term impacts are consistently considered. Additionally, incorporating intergenerational considerations into policy frameworks can be achieved by evaluating the distribution of costs, benefits, and risks across generations. This analysis can help identify potential imbalances and inform more equitable approaches to resource allocation and environmental management.

Engaging youth and future generations in sustainability efforts is also crucial. Educational programmes and initiatives that raise awareness about intergenerational equity can foster a culture of sustainability from a young age. By involving younger generations in environmental stewardship activities and decision-making processes, we can build a more informed and active constituency that advocates for long-term sustainability goals. For example, programmes that teach children about local ecosystems and the importance of conservation can instill a sense of responsibility and connection to the environment.

Furthermore, fostering a culture of sustainability through continuous public engagement and education is essential. Public campaigns that emphasise the ethical and practical importance of intergenerational equity can shift societal norms towards more sustainable consumption patterns and lifestyles. Encouraging dialogue between different generations can also bridge understanding and cooperation, ensuring that diverse perspectives are considered in planning and implementation.

Examples of Sustainable Development in Practice

Illustrating successful sustainable development projects around the world can serve as powerful examples to inspire and guide local initiatives. These case studies provide concrete evidence of how principles of Environmental Impact Assessment (EIA) can be integrated into project planning and implementation, demonstrating the real-world application of sustainability concepts.

Effective stakeholder engagement is a cornerstone of successful, sustainable development. For instance, in Kenya, the Olkaria Geothermal Project has demonstrated exemplary stakeholder involvement. The project incorporated extensive consultations with local communities, environmental groups, and government agencies from its inception. By actively involving stakeholders in the decision-making process, the project not only gained public support but also addressed potential conflicts early on. This collaborative approach ensured that the needs and concerns of diverse groups were considered, leading to more sustainable and accepted outcomes.

Another critical aspect of sustainable development is resource management strategies. The Suzhou Industrial Park in China exemplifies this by integrating cutting-edge resource management practices. The park employs a closed-loop system for water and waste management, significantly reducing consumption and environmental impact. Renewable energy sources, such as solar and wind power, are extensively used to meet energy demands. Additionally, the park promotes green building standards and energy-efficient technologies among its industries and residents, fostering a culture of sustainability within the community. Such comprehensive resource management strategies highlight the potential for scalable solutions that can be adapted to different contexts.

Adaptability and innovation also play a crucial role in the success of sustainable development projects. The Copenhagen Climate Plan stands as an excellent example of how projects can adapt to evolving circumstances and incorporate innovative solutions. Initially focused on reducing carbon emissions, the plan has progressively included new technologies and approaches, such as smart grid systems and urban farming, to address emerging environmental challenges. Copenhagen's ability to continuously innovate and modify its strategies in response to new data and changing conditions demonstrates the importance of flexibility in achieving long-term sustainability goals.

Measurable outcomes are essential for tracking progress and ensuring continuous improvement in sustainable development projects. The C40 Cities Climate Leadership Group, a network of the world's megacities committed to addressing climate change, utilises robust metrics to evaluate the effectiveness of their initiatives. By setting clear, measurable targets and regularly monitoring performance, these cities can identify areas for improvement and scale successful practices across the network. The use of performance indicators helps maintain transparency and accountability, ensuring that sustainability efforts remain on track and deliver tangible benefits to communities.

Summary and Reflections

This chapter has underscored the importance of integrating sustainability into project planning and implementation through the principles of Environmental Impact Assessment (EIA). By emphasising a comprehensive approach that considers economic, social, and environmental impacts collectively, EIAs can guide projects towards more sustainable outcomes. The discussion highlighted key principles such as inclusiveness, resilience, and the alignment with Sustainable Development Goals (SDGs), which provide a global framework for addressing various sustainability challenges.

Through practical examples, the chapter demonstrated how real-world applications of these principles can lead to significant benefits. Projects that incorporate green infrastructure or renewable energy not only mitigate environmental impacts but also offer economic and social advantages. The holistic approach advocated in this chapter ensures that development activities contribute to long-term sustainability, benefiting both current and future generations.

Reference List

Bernat, G. B., Qualharini, E. L., Castro, M. S., Barcaui, A. B., & Soares, R. R. (2023, January 1). Sustainability in Project Management and Project Success with Virtual Teams: A Quantitative Analysis Considering Stakeholder Engagement and Knowledge Management. Sustainability; mdpi. https://doi.org/10.3390/su15129834

Intergenerational Equity: An an overview | ScienceDirect Topics. (n.d.). Www.sciencedirect.com. https://www.sciencedirect.com/topics/social-sciences/intergenerational-equity

Oluwaseyi, J., & Stilinski, D. (2024). Promoting intergenerational equity in sustainable development: Ensuring future generations' needs through ethical and sustainable practices. ResearchGate. https://www.researchgate.net/publication/381306616_Promoting_Intergenerational_Equity_in_Sustainable_Development_Ensuring_Future_Generations'_Needs_through_Ethical_and_Sustainable_Practices

Precautionary Principles: An an overview | ScienceDirect Topics. (2013). Sciencedirect.com. https://www.sciencedirect.com/topics/earth-and-planetary-sciences/precautionary-principle

Pinto-Bazurco, J. F. (2020, October 23). The Precautionary Principle. International Institute for

Sustainable Development. https://www.iisd.org/articles/deep-dive/precautionary-principle

(PDF) Stakeholder Engagement: Achieving Sustainability in the Construction Sector. (n.d.). ResearchGate. https://www.researchgate.net/publication/275769628_Stakeholder_Engagement_Achieving_Sustainability_in_the_Construction_Sector

Polluter-Pays Principle: An an overview | ScienceDirect Topics. (n.d.). Www.sciencedirect.com. https://www.sciencedirect.com/topics/earth-and-planetary-sciences/polluter-pays-principle

Polluter Pays Principle: An an overview | ScienceDirect Topics. (2010). Sciencedirect.com. https://www.sciencedirect.com/topics/agricultural-and-biological-sciences/polluter-pays-principle

SDG Services. (n.d.). PRINCIPLES. SDG.SERVICES. https://www.sdg.services/principles.html

United Nations. (2024). The 17 Sustainable Development Goals. United Nations; United Nations. https://sdgs.un.org/goals

Chapter 9

Sector-Specific EIA Approaches

Evaluating Environmental Impact Assessments (EIAs) is essential for understanding and mitigating the environmental effects of various industry sectors. Each sector faces unique challenges that require tailored approaches to manage environmental concerns effectively. This chapter delves into the specific nuances and methodologies of EIAs across different industries, providing a comprehensive overview of the strategies employed to promote sustainable practices.

The chapter covers several critical areas, including the environmental impacts of mining projects, urban development and planning, coastal habitat protection, sustainable business practices, and water contamination issues. By exploring these topics, readers will gain insight into how EIAs are adapted to address the distinct environmental challenges associated with each sector. Practical examples and case studies illustrate the importance of informed decision-making and highlight the benefits of integrating advanced tools and technologies, such as Geographic Information Systems (GIS) and remote sensing, into the EIA process. This chapter aims not only to present the specific EIA approaches but also to underscore their significance in fostering sustainable development and natural resource management.

Mining Projects and Their Impacts

Mining activities have a significant impact on the environment, necessitating detailed Environmental Impact Assessments (EIAs) to mitigate adverse effects and promote sustainable practices. This section focusses on the specific environmental impacts of mining, including habitat destruction, water contamination, socio-economic effects, and regulatory compliance.

Mining often leads to habitat destruction, significantly affecting local flora and fauna. The clearing of vegetation for open-pit mining not only removes plant life but also disrupts animal habitats. For instance, in Ghana, large-scale gold mining has led to extensive deforestation, leading to the loss of biodiversity and disruption of ecosystems (Emmanuel et al., 2018). To address these issues, best practices for EIAs include baseline studies to identify existing species and their habitats. These studies should guide the development of conservation strategies such as setting aside protected areas or creating wildlife corridors to maintain ecological balance.

Understanding the pathways of contamination can inform regulatory measures for water pollution, which is another critical aspect of mining-induced environmental impact. Mining operations often involve the use of harmful chemicals like cyanide and mercury, which can leach into groundwater and surface water bodies. In many cases, tailings and waste rocks are sources of acid mine drainage, which leads to the contamination of local water systems with heavy metals and other pollutants (Haddaway et al., 2019). Effective EIAs must incorporate hydrological studies to map out potential contamination pathways. By identifying these routes, EIAs can recommend appropriate containment measures such as liners for tailing ponds, treatment facilities for effluents, and regular monitoring programmes to detect early signs of pollution.

Assessing the potential for displacement and contributing socio-economic indicators is critical for local communities. Mining projects often require land acquisition, leading to the displacement of people and the disruption of local livelihoods. For example, communities in mining areas of Ghana have experienced significant social upheaval, including the loss of agricultural lands and reduced access to resources necessary for traditional practices (Emmanuel et al., 2018). Best practices for EIAs should include comprehensive social impact assessments (SIAs) to evaluate the extent of displacement and its socio-economic consequences. These assessments should propose mitigation measures such as fair compensation, resettlement plans, and support for alternative livelihood programmes to ensure that affected communities are not left worse off.

Understanding local and international regulations can prevent legal challenges, which is essential for the smooth operation of mining projects. Mining activities are governed by a complex web of laws and regulations aimed at minimising environmental and social impacts. These regulations may vary significantly between countries and regions, requiring companies to stay abreast of both local and international standards. Failure to comply with these regulations can result in legal disputes, project delays, and financial penalties (Haddaway et al., 2019). Thus, robust EIAs must include a thorough review of relevant legislation and guidelines. This ensures that all project activities are planned and executed within the legal framework, thereby avoiding potential conflicts and fostering better relationships with regulatory bodies and stakeholders.

To achieve effective EIAs in the mining sector, several guidelines should be followed:

1. **Habitat Destruction** :

- Conduct comprehensive baseline environmental assessments to document existing flora and fauna.
- Develop and implement biodiversity management plans that include measures to protect critical habitats and endangered species.
- Consider creating buffer zones and wildlife corridors to minimise habitat fragmentation.

1. **Water contamination issues** :

- Map out contamination pathways through detailed hydrological studies.
- Implement containment measures such as lined tailing ponds and advanced wastewater treatment facilities.

- Establish continuous monitoring programmes to detect early signs of water contamination and take corrective action promptly.

1. **Socio-Economic Effects** :

- Perform in-depth social impact assessments to understand the extent of displacement and its socio-economic implications.
- Design and execute fair compensation and resettlement plans that align with best practice standards.
- Initiate community engagement programmes to involve local populations in decision-making processes, ensuring that their voices are heard and concerns addressed.

1. **Regulatory Compliance** :

- Stay informed about local and international mining regulations to ensure full compliance.
- Engage with regulatory authorities early in the project planning stage to clarify requirements and expectations.

- Develop and maintain comprehensive documentation to demonstrate adherence to legal and regulatory standards.

By implementing these guidelines, mining companies can significantly reduce environmental and social impacts while ensuring compliance with regulatory frameworks. The goal is to achieve a balance between resource extraction and environmental stewardship, ultimately contributing to sustainable development.

Urban Development and Planning

Enhancing urban planning processes through Environmental Impact Assessments (EIAs) is critical to integrating sustainability into city development. A well-executed EIA can inform and improve various facets of urban planning, from zoning to infrastructure design, ensuring that urban growth supports environmental, social, and economic objectives.

Awareness of zoning impacts plays a profound role in guiding responsible planning decisions. Zoning laws dictate land use patterns, which influence factors such as population density, green space allocation, and transportation networks. By incorporating EIAs into the zoning process, planners can evaluate the potential environmental effects of different zoning scenarios. For instance, higher-density residential zones may reduce urban sprawl but could increase local pollution levels if not managed properly. Conversely, commercial zoning clustered near public transit hubs can decrease reliance on personal vehicles, lowering emissions. These insights allow for zoning decisions that balance development needs with environmental sustainability, leading to more livable and resilient urban areas.

Evaluating the lifecycle impacts of infrastructure projects is another essential aspect of sustainable urban planning. Infrastructure has long-term environmental ramifications, from construction through demolition. By assessing these lifecycle impacts through EIAs, cities can make informed choices about materials, technologies, and designs that minimise negative environmental footprints. For example, using recycled materials or energy-efficient construction techniques can significantly reduce the carbon footprint of new buildings. Furthermore, evaluating the maintenance and operational phases ensures that infrastructure remains sustainable throughout its lifespan. This comprehensive approach fosters better design choices that align with sustainability goals, promoting a healthier urban environment.

Community engagement through participatory approaches is crucial for successful EIAs in urban planning. Involving local residents, businesses, and other stakeholders in the EIA process can foster trust and leverage valuable local knowledge. Participatory methods can include public consultations, workshops, and community surveys, providing forums where stakeholders can voice concerns and contribute to decision-making. This inclusive approach not only helps identify potential environmental issues early but also builds community buy-in for urban projects. When stakeholders see their input reflected in planning outcomes, they are more likely to support and collaborate on initiatives, enhancing the overall effectiveness and legitimacy of EIAs.

Highlighting case studies where ecosystem service valuation transformed planning strategies can serve as powerful examples to inspire new practices. Ecosystem services—such as air quality regulation, water filtration, and recreational spaces—are vital for urban areas. Valuing these services within EIAs can reveal the true cost and benefits of development projects, guiding more holistic decision-making. For example, in the Beijing-Tianjin-Hebei District, targeted adjustments in socioeconomic development based on ecosystem service evaluations led to significant improvements in water ecosystem health (Academic Press, 2016). Such case studies underscore the importance of integrating ecosystem service considerations into urban planning, demonstrating tangible benefits, and inspiring innovative planning strategies.

One notable example comes from Mount Carmel, Israel, where ecosystem service valuation proved instrumental in land management decisions. Here, researchers assessed changes in ecosystem services due to land-cover alterations, showing substantial gains in food provisioning and water regulation services despite challenges like wildfires and droughts (Campioli et al., 2016). Integrating such valuation techniques into urban EIAs can help cities recognise and preserve the ecological functions that underpin human well-being, driving more sustainable urban development.

The integration of life cycle assessment (LCA) and ecosystem services assessment (ESA) into EIAs offers a robust framework for evaluating urban projects' environmental impacts comprehensively. The LCA-ESA approach enables planners to account for both global and local environmental impacts, considering aspects like human health, natural systems, and resource use throughout a project's life cycle (Schaubroeck et al., 2016). By capturing a wide range of environmental interactions, this integrated method supports more informed and sustainable urban planning decisions.

For instance, examining wood production's lifecycle impacts as a provisioning service can guide sustainable forestry practices within urban green spaces. Similarly, assessing regulating services, such as CO_2 sequestration and water purification, provides insights into the broader environmental benefits of urban greenery. These assessments ensure that development projects enhance rather than undermine ecosystem services, aligning urban growth with sustainability objectives.

Coastal Habitat Protection

Understanding the fragility of ecosystems like mangroves and wetlands is critical for protection. These coastal habitats serve as vital buffers against storm surges, provide breeding grounds for marine life, and sequester significant amounts of carbon dioxide. Conducting Environmental Impact Assessments (EIAs) in these areas demands a thorough understanding of their ecological functions and vulnerabilities. Mangroves, for example, are highly sensitive to changes in salinity and water levels, making them particularly susceptible to environmental disturbances. Wetlands, similarly, host a diverse array of species that rely on stable water conditions. When planning projects in these regions, it's imperative to account for both the direct and indirect effects on these ecosystems.

Riding the line of adapting EIA practices to include climate resilience measures is becoming essential due to climate change. Coastal regions are often the first to feel the impacts of rising sea levels, increased storm frequency, and other climate-related challenges. Incorporating climate resilience into EIAs means evaluating how proposed projects will withstand these changes over time and what adaptations might be necessary. For instance, infrastructure projects near the coast should consider future sea level rise projections and incorporate designs that elevate or fortify structures accordingly. Additionally, restoring natural barriers such as dunes and mangroves can enhance a region's ability to absorb climatic shocks, providing dual benefits of habitat preservation and human protection.

Implementing guidelines for sustainable tourism can benefit both the economy and ecology. Coastal regions are popular tourist destinations, but unmanaged tourism can lead to significant environmental degradation. Sustainable tourism guidelines within EIAs ensure that development projects do not compromise the health of coastal ecosystems. This includes measures such as controlling visitor numbers to sensitive areas, promoting eco-friendly accommodation options, and educating tourists on the importance of preserving local habitats. By balancing tourist activities with conservation efforts, it is possible to create an economic model that supports both the community and the environment. For example, establishing marine protected areas can boost fish populations, subsequently enhancing opportunities for eco-tourism like snorkelling and scuba diving while protecting biodiversity.

Promoting regional collaboration highlights the need for united efforts across jurisdictions.

Environmental issues do not adhere to political boundaries; hence, collaboration among neighbouring regions is crucial for effective coastal management. Joint EIA initiatives can lead to more comprehensive understanding and solutions to shared environmental challenges. Cross-jurisdictional cooperation can facilitate data sharing, synchronised conservation efforts, and unified regulatory standards, thus ensuring that large-scale projects do not negatively impact adjacent areas. For instance, a regional approach to managing transboundary water bodies like estuaries can help maintain water quality and protect aquatic life. It also fosters a shared sense of responsibility and accountability among stakeholders, leading to more robust and coordinated environmental strategies.

Guidelines for conducting EIAs in coastal regions must address the specific vulnerabilities of these ecosystems. For mangroves and wetlands, it is essential to include assessments of hydrological regimes and potential alterations due to project activities. Climate resilience guidelines should emphasise adaptive management practices that can be modified as climate projections evolve. Sustainable tourism guidelines should encourage practices that minimise ecological footprints, such as promoting off-peak travel times and supporting local conservation projects. Regional collaboration guidelines should outline mechanisms for cooperative monitoring, enforcement, and dispute resolution to ensure cohesive environmental governance.

Understanding the interconnectedness of these factors is integral to developing effective EIAs for coastal regions. Each project presents unique challenges, requiring tailored approaches that consider both immediate and long-term environmental impacts. Effective protection measures stem from a deep appreciation of the delicate balance within these ecosystems and a commitment to preserving their integrity for future generations. The integration of climate resilience, sustainable tourism, and regional collaboration into EIA practices represents a forward-thinking approach to environmental stewardship, ensuring that coastal development proceeds in harmony with nature's needs.

The success of these initiatives relies heavily on the continuous engagement and education of all stakeholders involved. Local communities, government agencies, developers, and scientists must work together to share knowledge, resources, and responsibilities. Public awareness campaigns and stakeholder consultations can play a pivotal role in fostering such collaborations. By keeping the lines of communication open and transparent, it becomes easier to identify potential conflicts early and develop mutually beneficial solutions.

Moreover, technological advancements offer new tools for enhancing the accuracy and efficiency of EIAs. Utilising Geographic Information Systems (GIS), remote sensing, and predictive modelling enables more precise monitoring and assessment of coastal ecosystems. These technologies can help map sensitive habitats, track changes over time, and forecast potential impacts of proposed projects under various scenarios. Case studies from regions that have successfully integrated such tools into their EIA processes can provide valuable insights and lessons for others to follow.

Sustainable Business Practices in Various Sectors

In the quest for sustainable business practices, Environmental Impact Assessments (EIAs) serve as indispensable tools for enterprises across various sectors. This section emphasises how businesses can implement sustainable practices through effective EIA processes, thereby promoting environmental stewardship.

Understanding the connection between sustainability goals and business success is vital for modern enterprises. Companies today must recognise that their operations impact not only their financial performance but also the well-being of communities and the planet. As an example, Patagonia integrates environmental conservation into its core values by using recycled materials and reducing waste. Their commitment to sustainability not only aids in environmental preservation but also fosters consumer loyalty and brand reputation. Adopting a holistic approach that balances economic, social, and environmental considerations creates long-term benefits, driving business success while ensuring a positive impact on society and nature.

Presenting methods for minimising resource consumption and waste provides actionable insights for businesses striving to enhance sustainability. One effective strategy is adopting a life cycle perspective, considering the entire lifespan of a product or service—from raw material extraction to disposal. For instance, Tesla assesses the environmental impact of battery production and end-of-life recycling in addition to producing zero-emission vehicles. Businesses can optimise resource use and minimise waste by designing products with longevity, using renewable energy sources, adopting energy-efficient technologies, and exploring eco-friendly alternatives. The circular economy model, which promotes recycling and resource efficiency, offers another robust framework. Interface, a carpet manufacturer, has successfully implemented this model by creating modular carpets that can be disassembled, recycled, and reassembled, thus conserving resources and reducing waste.

Illustrating how innovative assessment technologies improve EIAs encourages adoption in diverse sectors. Geographic Information Systems (GIS) and remote sensing are two powerful tools enhancing EIA effectiveness. These technologies allow for the comprehensive integration and analysis of various data sources, providing a clear environmental context. Advanced modelling and simulation techniques help predict potential impacts under different scenarios, leading to more informed decision-making. For example, AI algorithms can analyse large volumes of data, identifying patterns and trends that may not be evident to human analysts. Nestlé leverages these technologies to evaluate packaging choices' environmental implications, resulting in innovations like recyclable or biodegradable packaging materials. Embracing such technologies facilitates more accurate assessments and supports the development of targeted mitigation measures, thereby improving overall EIA outcomes.

Integrating Corporate Social Responsibility (CSR) into EIA processes promotes transparency and accountability, leading to better outcomes. CSR initiatives ensure that companies consider their broader impact on society and the environment. A practical example is Unilever's Sustainable Living Plan, which aims to reduce environmental impact, improve health and well-being, and enhance livelihoods by 2030. Incorporating EIA into CSR frameworks allows businesses to assess and mitigate their activities' social and environmental impacts systematically. This approach fosters stakeholder engagement and public participation, ensuring that community concerns and interests are addressed. Additionally, transparent reporting through sustainability reports and certifications, like LEED (Leadership in Energy and Environmental Design), demonstrates a company's commitment to responsible practices. Such transparency builds trust with stakeholders and enhances the company's reputation.

Effective EIA processes also play a crucial role in risk assessment and mitigation. By identifying potential environmental impacts, companies can develop strategies to minimise adverse effects. This involves comprehensive studies to evaluate possible impacts on air, water, soil, biodiversity, and ecosystems. For example, mining companies conduct EIAs to assess the environmental and social implications of their operations, enabling them to adopt practices that protect local ecosystems and communities. Similarly, construction firms evaluate the effects of building projects on nearby habitats and cultural heritage sites, ensuring that development activities sustain rather than harm local environments.

Moreover, integrating sustainability reporting and CSR with EIA provides a holistic view of corporate performance. Combining these practices highlights the interconnectedness of environmental, social, and economic factors, allowing for more comprehensive decision-making. It also offers a platform for continuous improvement through regular monitoring and feedback. Businesses can report progress on their sustainability goals, making necessary adjustments based on EIA findings. This integrated approach supports the alignment of business strategies with sustainability objectives, ensuring that companies remain accountable while advancing their sustainability agendas.

The potential for Artificial Intelligence (AI) and big data to enhance EIA cannot be overlooked. AI can automate data collection and analysis tasks, freeing up human resources for more strategic activities. Big data, including real-time monitoring and historical trends, provides a wealth of information for EIAs. These technologies enable more accurate impact predictions and the design of effective mitigation measures. By leveraging AI and big data, decision-makers can access comprehensive and up-to-date information, leading to more informed and effective decisions. Google's investment in renewable energy and energy storage solutions exemplifies how technology-driven EIAs can drive sustainable business practices.

Water Contamination Issues

Water contamination arising from industrial activities is a pressing environmental concern. Controlling and mitigating this issue effectively requires comprehensive Environmental Impact Assessments (EIAs). Industrial processes often release a variety of pollutants into water bodies, significantly affecting aquatic life, ecosystems, and human health. This section delves into the significance of such contamination and explores how EIAs can address these challenges.

Firstly, understanding the pathways of water contamination is crucial in shaping effective regulatory measures. Contaminants can enter water systems through direct discharge, surface runoff, or leaching from soils. For example, heavy metals such as lead, cadmium, and nickel are commonly found in industrial effluents and have been shown to exceed recommended values proposed by WHO and EPA (Afrad et al., 2020). These metals pose significant risks to human health and the environment. By identifying how these contaminants travel through different media, authorities can develop targeted regulations to minimise their impact.

The necessity of establishing strict effluent discharge standards cannot be overstated. Setting stringent limits on the concentration of harmful substances in industrial discharges is a critical step towards safeguarding water quality. Regulatory agencies must base these standards on scientific evidence and best practices. For instance, the study by Ashraf et al. demonstrates the effectiveness of the CIF-ELECTREE method in providing accurate assessments of industrial pollution (M. et al., 2023). Such advanced methods can inform the creation of robust discharge standards that mitigate contamination risks.

Case studies on successful clean-up actions illustrate the effectiveness of well-implemented EIAs. One notable example is the remediation of the Turag River in Bangladesh. Prior to intervention, the river suffered severe pollution due to indiscriminate industrial effluent discharge, which adversely affected water quality and surrounding ecosystems (Afrad et al., 2020). Through rigorous EIAs and subsequent clean-up efforts, including the implementation of stricter discharge regulations and regular monitoring, the water quality of the Turag River has seen marked improvements. These case studies serve as valuable references for other regions facing similar challenges.

Ensuring continuous monitoring of water quality is vital for compliance and future risk mitigation. Regular sampling and analysis of water sources help detect any deviations from established standards, enabling timely corrective actions. Advanced monitoring technologies, such as Geographic Information Systems (GIS) and remote sensing, enhance our ability to track water quality changes over time. In regions where industrial activity is prevalent, deploying continuous monitoring systems ensures that immediate responses can be initiated if contamination levels rise unexpectedly.

Moreover, communities and industries must collaborate for effective water quality management. Public awareness campaigns can educate local populations about the importance of maintaining clean water sources and the potential dangers of industrial pollution. Industries should adopt cleaner production techniques and invest in waste treatment facilities to reduce pollutant loads. Government incentives for industries that achieve and maintain high environmental standards can also drive positive change.

Implementing these strategies requires a coordinated effort among stakeholders. Governments must enforce regulations consistently, while industries need to comply with legal requirements and embrace sustainable practices. Community involvement ensures transparency and accountability, fostering a collective responsibility for protecting water resources.

Research plays an essential role in refining EIA processes and enhancing their effectiveness. Studies evaluating the impact of various pollutants on water quality provide critical insights for developing better assessment tools. For example, incorporating methodologies like the CIF-ELECTREE method helps to account for uncertainties and complexities in pollution data, leading to more reliable decision-making (M. et al., 2023).

As technology advances, integrating innovative solutions into EIA processes becomes increasingly feasible. New sensor technologies and data analytics platforms can offer real-time insights into water quality dynamics, improving the overall responsiveness of environmental management systems. These advancements facilitate more proactive interventions, reducing the likelihood of severe contamination events.

Insights and Implications

This chapter provided a comprehensive examination of Environmental Impact Assessments (EIAs) within the mining industry, addressing the critical environmental and socio-economic impacts associated with such projects. Key issues such as habitat destruction, water contamination, socio-economic disruptions, and regulatory compliance were discussed in detail. The analysis highlighted the need for baseline studies, hydrological assessments, and social impact assessments to develop effective mitigation strategies. By implementing best practices, including biodiversity management plans, advanced wastewater treatment facilities, fair compensation schemes, and aligning with local and international regulations, mining companies can significantly reduce their environmental and social footprint.

The guidelines outlined in this chapter aim to foster sustainable practices in the mining sector, balancing resource extraction with environmental stewardship. These recommendations focus on minimising adverse effects on ecosystems and communities while ensuring adherence to regulatory standards. Through continuous monitoring, community engagement, and collaboration with regulatory authorities, mining operations can achieve a more sustainable and responsible approach. Ultimately, the insights gained from EIAs not only help protect the environment but also contribute to the overall well-being of affected populations, promoting a harmonious coexistence between industrial activities and natural ecosystems.

Reference List

Afrad, M. S. I., Monir, M. B., Haque, M. E., Barau, A. A., & Haque, M. M. (2020, June 30). *Impact of industrial effluent on water, soil and Rice production in Bangladesh: a case of Turag River Bank* . Journal of Environmental Health Science and Engineering. https://doi.org/10.1007/s40201-020-00506-8

Schaubroeck et al. (2016). *Environmental impact assessment and monetary ecosystem service valuation of an ecosystem under different future environmental change and management scenarios: a case study of a Scots pine forest* . *Journal of Environmental Management* , 173, 79-94. https://doi.org/10.1016/j.jenvman.2016.03.005

Ashraf, S., Ali, M., Sohail, M., & Eldin, S. M. (2023). *Assessing the environmental impact of industrial pollution using the complex intuitionistic fuzzy ELECTREE method: a case study of pollution control measures* . *Frontiers in Environmental Science* , 11, Article 1171701. https://doi.org/10.3389/fenvs.2023.1171701

Editor. (2022, July 24). *The Role of Environmental Impact Assessments in Corporate Decision-Making* . Michael Edwards | Commercial Corporate Solicitor. https://michaeledwards.uk/the-role-of-environmental-impact-assessments-in-corporate-decision-making/

Emmanuel, A. Y., Jerry, C. S., & Dzigbodi, D. A. (2018, March). *Review of Environmental and Health Impacts of Mining in Ghana* . Journal of Health and Pollution. https://doi.org/10.5696/2156-9614-8.17.43

FasterCapital. (n.d.). *Sustainable business practices to reduce environmental impact* . Retrieved from https://fastercapital.com/topics/sustainable-business-practices-to-reduce-environmental-impact.html/5

Haddaway, N. R., Cooke, S. J., Lesser, P., Macura, B., Nilsson, A. E., Taylor, J. J., & Raito, K. (2019, February 21). *Evidence of the impacts of metal mining and the effectiveness of mining mitigation measures on social-ecological systems in Arctic and boreal regions: a systematic map protocol* . Environmental Evidence: https://doi.org/10.1186/s13750-019-0152-8

Schaubroeck, T., Deckmyn, G., Giot, O., Campioli, M., & Muys, B. (2016). *Environmental impact assessment and monetary ecosystem service valuation of an ecosystem under different future environmental change and management scenarios: A case study of a Scots pine forest* . *Journal of Environmental Management* , *173* , 79-94. https://doi.org/10.1016/j.jenvman.2016.03.005

Chapter 10

Future Trends and Innovations in EIA

Exploring future trends and innovations in Environmental Impact Assessment (EIA) is essential for advancing sustainable resource management practices. The constantly evolving field of EIA benefits greatly from emerging technologies and methodologies that promise to enhance the precision, efficiency, and effectiveness of environmental assessments. By understanding these advancements, practitioners and policymakers can better anticipate and mitigate environmental impacts, contributing to more informed decision-making processes.

This chapter delves into significant technological breakthroughs such as advancements in remote sensing and Geographic Information Systems (GIS), which offer improved data accuracy and collaboration opportunities. It examines the integration of high-resolution satellite imagery for detailed environmental monitoring and the deployment of cloud-based GIS platforms for real-time data sharing among stakeholders. Machine learning applications are explored for their potential to provide sophisticated data analysis and prediction capabilities, while drone technology is highlighted for its ability to gather data in challenging terrains. The part of artificial intelligence (AI) and how it has changed EIA processes is also talked about. This includes automated data collection, better risk assessment models, AI-powered public engagement tools, and learning from past assessments all the time. The chapter also addresses the growing importance of citizen science in EIAs, showcasing how public involvement can enrich data collection and foster transparency. Finally, it anticipates policy and regulatory changes that will shape the future of EIA practices, emphasising international cooperation, the inclusion of social and economic impact assessments, and the adaptation of regulations to accommodate technological innovations. Each section provides practical guidelines to help practitioners effectively integrate these new technologies and approaches into their EIA workflows.

Advancements in Remote Sensing and GIS

Remote sensing and Geographic Information Systems (GIS) are crucial components in Environmental Impact Assessments (EIA), offering precision and efficiency. Technological advancements in these areas have transformative potential, enhancing data accuracy and collaboration among stakeholders. This subpoint explores various technological enhancements that promise to revolutionise EIA processes.

The integration of high-resolution satellite imagery improves the accuracy of data for environmental monitoring. Satellite imagery provides a macroview of large geographical areas, which is essential for understanding environmental changes over time. Recent advancements have made it possible to capture images with higher resolution, enabling more detailed observation of land cover, vegetation, water bodies, and infrastructure. High-resolution images can detect minute changes in the environment, which is invaluable for monitoring deforestation, habitat loss, and other ecological disturbances. For instance, satellites can now monitor the health of forests by detecting variations in leaf area index, chlorophyll content, and other vegetative indicators (Quamar et al., 2023). By providing comprehensive and precise data, high-resolution imagery allows environmental professionals to make informed decisions based on up-to-date information.

The development of cloud-based GIS platforms enables better collaboration among practitioners. Traditional GIS software often required substantial computational resources, limiting its accessibility and collaborative potential. Cloud-based platforms, however, offer scalable computing power and storage, making GIS tools more affordable and easier to use. These platforms facilitate real-time data sharing and collaborative analysis, allowing multiple stakeholders to contribute and access up-to-date information simultaneously. Environmental professionals from different regions or departments can work together on a single project without geographic constraints. Additionally, cloud-based GIS systems support various data formats and integrate seamlessly with other digital tools, streamlining workflows and improving efficiency. Guidelines for implementing cloud-based GIS include ensuring data security through robust encryption methods and providing adequate training for users to maximise their proficiency and effectiveness.

Machine learning applications facilitate more sophisticated data analysis and prediction capabilities. Machine learning algorithms excel at recognising patterns and making predictions from large datasets, which are common in environmental studies. Integrating machine learning with GIS enhances the ability to analyse complex datasets, such as climate models, pollution levels, and biodiversity indices. For example, supervised machine learning models can predict the impact of urban development on local wildlife habitats by analysing historical data and identifying correlations between land-use changes and species distributions (Li et al., 2023). Unsupervised learning techniques, on the other hand, can uncover hidden patterns and anomalies in environmental datasets, providing insights that may not be apparent through traditional analytical methods. To effectively harness machine learning in GIS, practitioners should follow guidelines that include selecting appropriate algorithms based on the specific research question, validating models with independent datasets, and continuously updating models with new data to maintain accuracy.

Drones provide an efficient way to gather data in remote or difficult-to-access areas. Drone technology has advanced rapidly, offering new opportunities for environmental monitoring and data collection. Equipped with high-resolution cameras, LiDAR sensors, and multispectral imagers, drones can capture detailed spatial data from hard-to-reach locations, such as dense forests, mountainous regions, and wetlands. For instance, drones can be used to monitor coastal erosion, map flooding extents during natural disasters, and track wildlife movements with minimal disturbance. The high mobility and flexibility of drones allow for frequent and targeted data collection, covering large areas quickly and cost-effectively. Moreover, drone-collected data can be integrated into GIS platforms, enriching datasets with high-resolution, site-specific information. To optimise the use of drones in EIAs, practitioners should adhere to guidelines that include obtaining necessary flight permits, conducting risk assessments to ensure safe operations, calibrating sensors regularly, and adhering to ethical practices to minimise environmental impact.

Impact of AI and Machine Learning

Artificial intelligence and machine learning are revolutionising Environmental Impact Assessment (EIA) processes by enhancing the accuracy, efficiency, and predictive capabilities of environmental analyses. This subpoint explores how these technologies are being integrated into EIAs, transforming traditional methodologies into more automated, precise, and adaptive systems.

Automated data collection and analysis have significantly streamlined the data gathering process in EIAs. Traditional methods of data collection are often labour-intensive and time-consuming, involving manual sampling and analysis. However, AI-driven tools can automate these tasks, collecting data from various sources such as sensor networks, satellite imagery, and even social media. Then, machine learning algorithms analyze this data to find patterns and trends that human analysts might miss. For instance, AI can monitor air and water quality in real-time, providing continuous data streams that offer a more comprehensive understanding of environmental conditions. This level of automation not only reduces the workload for EIA practitioners but also enhances the accuracy and timeliness of the assessments.

Enhanced risk assessment models using AI are another vital advancement in the field of EIAs. Predicting potential environmental risks has always been a core component of EIAs, but traditional models often fall short due to their reliance on linear relationships and limited data sets. AI algorithms, particularly machine learning models, can handle vast amounts of complex, nonlinear data, making them exceptionally suited for this task. According to research, deploying machine learning techniques in soil pollution management has shown improved model performance in predicting environmental impacts (Elsevier, 2024). These models can anticipate potential risks by analysing historical data and identifying early warning signals, thus enabling more proactive mitigation measures. For example, AI models can predict the likelihood of harmful algal blooms by analysing parameters such as water temperature, nutrient levels, and historical bloom occurrences.

AI-driven public engagement tools are enhancing communication with stakeholders, an essential aspect of EIAs. Effective stakeholder engagement is crucial for the success of any EIA, as it ensures that the concerns and inputs of all affected parties are considered. Traditional methods of stakeholder engagement, such as public meetings and written submissions, can be limiting and may not reach a broad audience. AI can address these challenges by offering interactive and accessible platforms for public participation. Chatbots, for example, can provide instant responses to queries about the EIA process, while sentiment analysis tools can gauge public opinion from social media posts. Additionally, virtual reality (VR) applications can create immersive simulations of proposed projects, helping stakeholders visualise potential impacts more clearly. By facilitating better communication and feedback mechanisms, AI-driven tools ensure that stakeholder engagement is more inclusive and effective.

Another significant advantage of integrating AI in EIAs is continuous learning from previous assessments. Traditional EIA methodologies often rely on static models and assumptions that may not adapt well to changing environmental conditions. In contrast, machine learning systems can continuously update their algorithms based on new data and insights, leading to progressively better predictions over time. For instance, machine learning models used in previous EIAs can be retrained with the latest environmental data, improving their predictive accuracy for future assessments. This continuous learning capability ensures that EIA practices remain current and responsive to emerging environmental trends and challenges. The use of reinforcement learning techniques allows these models to learn from their mistakes and successes, further refining their predictive abilities.

Implementing AI and machine learning in EIAs also involves addressing certain challenges to maximise their potential benefits. One key challenge is ensuring the quality and reliability of the data used to train AI models. Data sparsity and variability can impact model performance, making it essential to adopt robust data pre-processing and validation techniques. As highlighted in recent studies, log-normalization and other data transformation methods can enhance the predictive capability of machine learning models (Elsevier, 2024). Additionally, integrating data from diverse sources, such as remote sensors, historical records, and community inputs, can help build more comprehensive and accurate models.

Another challenge is the transparency and interpretability of AI models. Stakeholders and decision-makers need to understand how AI models arrive at their predictions and recommendations. This requires developing explainable AI techniques that can elucidate the decision-making processes of complex algorithms. Tools such as SHAP (Shapley Additive Explanations) and LIME (Local Interpretable Model-Agnostic Explanations) can provide insights into model behaviour, ensuring that AI-driven decisions are transparent and trustworthy.

Moreover, the successful integration of AI in EIAs necessitates building capacity among practitioners. Training and education programmes are essential to equip environmental professionals with the skills needed to utilise AI tools effectively. This includes understanding the principles of machine learning, data science, and algorithmic development. Collaboration between AI experts and environmental scientists can foster interdisciplinary approaches, combining domain knowledge with technical expertise to solve complex environmental problems.

The ethical implications of using AI in EIAs must also be considered. Ensuring that AI applications adhere to principles of fairness, accountability, and inclusivity is paramount. This includes addressing potential biases in AI models, protecting data privacy, and ensuring equitable access to AI tools across different communities and regions. Policymakers and regulators play a crucial role in establishing guidelines and standards that govern the ethical use of AI in environmental assessments.

Role of Citizen Science

Citizen science, a concept that integrates public involvement in scientific research, has been making significant strides in the field of Environmental Impact Assessment (EIA). This subpoint elucidates the growing importance of citizen science in EIA, emphasising how community engagement can offer valuable data and insights into assessing environmental impacts.

Involving local communities in data collection for EIA empowers participants while enriching the assessment process. When community members are engaged in gathering environmental data, they develop a sense of ownership and responsibility towards their environment. This empowerment not only enhances the quality of data collected but also cultivates a more informed and environmentally conscious populace. For instance, in coastal regions, local fishermen participating in monitoring water quality provide real-time data on pollution levels, which is vital for timely interventions. Encouraging community participation fosters a collaborative spirit, ensuring diverse perspectives and observations are included in the assessment process.

Crowdsourced monitoring initiatives represent another hallmark of citizen science in EIA. These initiatives expand data collection efforts beyond traditional methodologies, allowing for more extensive and frequent environmental monitoring. By leveraging the collective power of the public, large volumes of data can be gathered quickly and cost-effectively. Mobile applications and online platforms enable citizens to report environmental observations, such as deforestation, wildlife sightings, or water contamination. One prominent example is the Naturalist platform, where users worldwide share biodiversity data, significantly contributing to global databases used for ecological studies and assessments. Such initiatives enhance the granularity and scope of environmental data, improving the accuracy and comprehensiveness of EIAs.

Enhancing transparency and trust through citizen engagement is crucial for improving public perception and accountability in EIA processes. When communities are actively involved in environmental assessments, they gain direct insights into the methodologies and findings, fostering a transparent evaluation process. This openness helps build trust between stakeholders, including policymakers, scientists, and the public. Transparent practices ensure that the data collected and the subsequent assessments are viewed as credible and legitimate. For example, in air quality monitoring projects, providing community members with access to real-time data dashboards allows them to verify results independently, increasing confidence in the findings and recommendations derived from the EIA.

Several case studies illustrate the effective collaboration between professionals and citizens in successful citizen science projects. One notable example is the Extreme Citizen Science project, which employs participatory methods to address local environmental issues. In this initiative, residents of various communities gather traditional ecological knowledge alongside modern scientific data to document changes in their environment. Tools like Sapelli and Tap facilitate data collection even in remote areas, empowering local populations to contribute meaningfully. These projects have demonstrated significant social and environmental benefits, showcasing how community engagement can translate into actionable outcomes (Eglė Butkevičienė et al., 2021).

Another example is the "Can I breathe?" initiative in the Czech Republic, where citizens participate in monitoring air quality using open-source software and portable sensors. This project not only raises awareness about air pollution but also equips residents with the tools to advocate for policy changes based on empirical evidence. Similarly, the Fortepan project in Hungary exemplifies how historical environmental data can be crowdsourced to create a comprehensive archive of ecological changes over time, aiding current and future assessments.

Guidelines for involving local communities in data collection emphasise the need for structured training programmes to educate participants on data collection techniques and tools. Providing clear instructions and resources ensures the reliability and consistency of the data collected. Additionally, establishing feedback mechanisms where participants receive updates on how their contributions impact the assessment process can maintain motivation and sustained engagement.

For crowdsourced monitoring initiatives to be successful, it is essential to integrate user-friendly technologies that simplify data reporting. Developing mobile applications with intuitive interfaces encourages broader participation. Ensuring data privacy and security is also paramount to maintaining public trust. Implementing robust verification protocols can help validate the authenticity of the crowdsourced data, enhancing its credibility for inclusion in EIAs (Public Engagement: Search | Citizen Science: Theory and Practice, 2023).

Enhancing transparency and trust through citizen engagement involves regular communication and the dissemination of findings to the public. Hosting community meetings where preliminary results and methodologies are presented allows for open discussion and feedback. Providing access to raw data and analytical processes further demonstrates a commitment to transparency. Establishing independent review panels comprising community representatives and experts can oversee the EIA processes, reinforcing accountability.

The influence of citizen science on EIA extends beyond data collection to fostering a participatory culture in environmental governance. By involving citizens directly, these projects encourage democratic decision-making processes where public opinion and local knowledge play pivotal roles. This inclusive approach ensures that environmental policies and actions reflect the needs and values of the communities they affect.

Policy and Regulatory Changes on the Horizon

Anticipated changes in policies and regulations are set to significantly shape Environmental Impact Assessment (EIA) practices and natural resource management. The need to address urgent environmental issues, particularly climate change, is what has sparked these changes. One of the most critical areas where we expect significant policy evolution is the increased emphasis on sustainability in regulations. This shift focusses on integrating long-term environmental stewardship into regulatory frameworks to better address climate change impacts. For instance, regulations are likely to increasingly mandate that EIA processes consider not only immediate environmental impacts but also long-term sustainability goals. This would ensure that projects contribute positively to mitigating climate change effects, reducing greenhouse gas emissions, and promoting renewable energy use.

Future collaboration between countries on environmental assessments will enhance consistency and cooperation. As environmental issues often transcend national borders, international cooperation becomes vital. Harmonising EIA standards across countries can lead to more consistent and effective environmental protection efforts. By working together, nations can share best practices, data, and technological advancements, ensuring that EIAs are conducted uniformly and effectively worldwide. Collaborative efforts could also involve joint environmental monitoring programmes and shared databases, which would facilitate more comprehensive and accurate assessments. Such initiatives would foster a global approach to environmental protection, with countries supporting each other in achieving sustainable development goals.

Incorporating social and economic impact assessments into EIAs is another anticipated regulatory change that promises to improve decision-making processes. Traditionally, EIAs have primarily focused on biophysical impacts, often overlooking the social and economic dimensions. However, integrating these aspects ensures a more holistic evaluation of proposed projects. Social impact assessments (SIAs) examine how projects affect communities, considering factors such as displacement, employment opportunities, and public health. Economic impact assessments (EIAs) evaluate the financial implications for local economies, including cost-benefit analyses and potential market effects. By combining environmental, social, and economic assessments, decision-makers can gain a fuller understanding of a project's overall impact, leading to more informed and equitable choices.

The adaptation of regulations to technological innovations is essential for the rapid integration of new methodologies into EIA practices. Technological advancements offer numerous opportunities for improving the efficiency and accuracy of EIAs. For example, remote sensing technologies and Geographic Information Systems (GIS) provide precise and real-time environmental data, enabling more detailed and timely assessments. Regulations must evolve to accommodate these innovations, ensuring they are incorporated seamlessly into the EIA process. This includes updating guidelines to reflect current technological capabilities and providing training for practitioners to effectively utilise new tools. Moreover, fostering an adaptable regulatory environment encourages continuous improvement and innovation within the field.

Guidelines play a crucial role in facilitating collaboration between countries on environmental assessments. Standardised procedures help ensure that all participating nations follow consistent methodologies, leading to comparable and reliable results. Guidelines should outline key principles for collaborative assessments, such as data sharing protocols, joint monitoring frameworks, and mechanisms for resolving disputes. Establishing clear communication channels and periodic review processes further enhances coordination among countries. By adhering to robust guidelines, nations can collectively address transboundary environmental issues more effectively.

The inclusion of social and economic impact assessments in EIAs requires structured guidelines to standardise these evaluations. Such guidelines should detail the necessary steps and criteria for assessing social and economic impacts, ensuring a comprehensive and unbiased analysis. This involves identifying affected stakeholders, gathering relevant data, and conducting thorough analyses of the potential consequences. Clear criteria for evaluating both positive and negative impacts enable balanced and transparent decision-making. Additionally, guidelines should emphasise stakeholder engagement throughout the assessment process, foster dialogue, and ensure that community voices are heard and considered.

Adapting regulations to accommodate technological innovations also necessitates updated guidelines. These guidelines should focus on the integration of new tools and methods into existing EIA frameworks, ensuring that advancements are utilised effectively and responsibly. Key components might include criteria for validating new technologies, protocols for data collection and analysis, and standards for reporting results. Training modules and certification programmes can also be part of these guidelines, equipping practitioners with the skills needed to harness technological advancements. By maintaining up-to-date guidelines, regulatory bodies can ensure that EIA practices remain cutting-edge and capable of addressing contemporary environmental challenges.

Increased emphasis on sustainability in regulations has far-reaching implications for climate change mitigation. Policies that prioritise sustainability encourage the adoption of green technologies and practices, thereby reducing carbon footprints and enhancing ecosystem resilience. For example, regulations may incentivise the use of renewable energy sources or mandate energy efficiency measures in development projects. Such policies not only contribute to environmental preservation but also promote economic benefits through the creation of green jobs and the stimulation of sustainable industries. By embedding sustainability into regulatory frameworks, governments can drive systemic changes that align development activities with broader climate goals.

Future collaboration between countries on environmental assessments is essential for tackling global environmental issues. International partnerships can pool resources and expertise, enabling more comprehensive and effective EIAs. Strategies for fostering collaboration include establishing multinational working groups to develop joint assessment protocols and conducting cross-border environmental studies. Furthermore, international treaties and agreements can formalise commitments to collaborative environmental governance. By working together, countries can address shared environmental challenges, such as air and water pollution, habitat destruction, and climate change, more effectively.

The incorporation of social and economic impact assessments leads to well-rounded decision-making, enhancing the credibility and acceptability of EIA outcomes. Considering the social and economic dimensions of projects ensures that their benefits and costs are distributed fairly among different societal groups. This approach helps to prevent adverse social effects, such as displacement or loss of livelihoods, and maximises positive economic impacts, such as job creation and community development. Transparent and inclusive assessment processes build trust among stakeholders, fostering public support and compliance with environmental regulations.

Adapting regulations to technological innovations enables the rapid deployment of advanced methodologies within EIA processes. Technologies such as machine learning, big data analytics, and drone surveillance offer unprecedented opportunities for enhancing the scope, accuracy, and efficiency of EIAs. Regulations must be flexible and forward-looking to accommodate these advancements, ensuring they can be effectively integrated into practice. This involves continuously reviewing and updating regulatory frameworks to reflect technological progress and promote research and development in environmental technologies. By staying ahead of technological trends, EIA practices can remain relevant and robust in the face of evolving environmental challenges.

Technological Innovations Improving EIA Efficiency

The integration of new technologies into Environmental Impact Assessments (EIA) holds the promise of significant improvements in process efficiency and outcome quality. This subpoint delves into specific technological advancements and their transformative capabilities within EIA.

One prominent advancement is the use of advanced sensors, which significantly enhance the precision of environmental data collection. Traditional methods often rely on manual data gathering, which can be time-consuming and subject to human error. Advanced sensors, however, provide continuous and real-time monitoring, collecting extensive data sets with high accuracy. These sensors can measure various parameters, such as air and water quality, soil conditions, and wildlife activity. Such precise data allows for more detailed and accurate environmental impact evaluations. For instance, sensors deployed in marine environments can monitor changes in water temperature and pollution levels, providing critical data to assess the impact of offshore drilling activities.

Real-time data analytics bolster this by improving the speed and relevance of assessments. In traditional EIA processes, data collected from field surveys often undergoes lengthy analysis phases, delaying decision-making. Real-time data analytics allows for the immediate processing and analysis of data gathered by sensors and other tools. This means that potential environmental impacts can be identified and addressed much quicker. Real-time analytics also allow for adaptive management strategies where mitigation measures can be adjusted dynamically based on the most current data. For example, construction projects near protected areas can continuously monitor noise levels and adjust activities to minimise disruption to local wildlife in real time.

Virtual reality (VR) tools represent another significant innovation. VR provides immersive visualisations that are invaluable for stakeholder engagement and education. Traditional EIA reports, often composed of dense scientific data and descriptive text, can be challenging for non-experts to interpret. VR overcomes this by offering interactive 3D models and simulations of proposed projects and their potential impacts on the environment. Stakeholders, including community members, policymakers, and project developers, can virtually explore scenarios such as the construction of a new wind farm or the restoration of a wetland. This experience facilitates better understanding and more meaningful discussions about the potential environmental consequences and mitigation strategies.

Blockchain technology offers a revolutionary approach to ensuring secure and transparent data management within EIA processes. One of the key challenges in environmental assessments is maintaining the integrity and transparency of data throughout the project's lifecycle. Blockchain addresses this challenge by providing a decentralised and immutable ledger for recording and storing data. Every data entry or change is time-stamped and cannot be altered without consensus from the network. This level of security ensures that the data used in EIAs is accurate, verifiable, and resistant to tampering. For instance, data on pollutant emissions from industrial sites recorded on a blockchain can be reliably tracked and audited, enhancing regulatory compliance and public trust.

Integrating these technologies into EIA processes not only optimises the efficiency and accuracy of assessments but also fosters more sustainable resource management practices. The use of advanced sensors for precise data collection allows for more comprehensive environmental monitoring. Real-time analytics accelerate the assessment process, enabling quicker responses to potential environmental issues. VR tools enhance communication and understanding among stakeholders, leading to more informed and collaborative decision-making. Finally, blockchain technology secures and validates data, ensuring transparency and accountability.

Final Thoughts

This chapter has examined the latest advancements in Environmental Impact Assessment (EIA), highlighting how emerging technologies and methodologies are set to transform the field. By leveraging tools such as high-resolution satellite imagery, cloud-based GIS platforms, machine learning applications, and drones, environmental professionals can achieve greater precision and efficiency in data collection and analysis. These innovations facilitate better collaboration among stakeholders and enable more detailed monitoring of environmental changes, leading to more informed decision-making.

The integration of artificial intelligence (AI) and citizen science further enhances EIA practices by introducing automation, predictive capabilities, and increased public engagement. AI-driven models streamline data processing and improve risk assessments, while citizen science initiatives empower local communities to contribute valuable data. Policy and regulatory changes are also shaping the future of EIAs by emphasising sustainability, international cooperation, and the incorporation of social and economic impact assessments. Together, these technological and methodological advancements promise a more comprehensive and effective approach to managing environmental impacts, ultimately contributing to more sustainable resource management practices.

Reference List

Anifowose(2024). Artificial intelligence and machine learning in environmental impact prediction for soil pollution management – case for EIA process. Environmental Advances, 17, 100554. https://doi.org/10.1016/j.envadv.2024.100554

Eglė Butkevičienė, Artemis Skarlatidou, Bálint Balázs, Barbora Duží, Massetti, L., Ioannis Tsampoulatidis, & Tauginienė, L. (2021, January 1). Citizen Science Case Studies and Their Impacts on Social Innovation. Springer eBooks; Springer Nature. https://doi.org/10.1007/978-3-030-58278-4_16

EIA Guidelines for Assessing the Impact of Climate Change on a Project | Sabin Centre for Climate Change Law. (n.d.). Climate.law.columbia.edu. https://climate.law.columbia.edu/content/eia-guidelines-assessing-impact-climate-change-project

Javaid, M., Haleem, A., Singh, R. P., Suman, R., & Gonzalez, E. S. (2022, January). Understanding the Adoption of Industry 4.0 Technologies in Improving Environmental Sustainability. Sustainable Operations and Computers; ScienceDirect. https://doi.org/10.1016/j.susoc.2022.01.008

Li, F., Tan Yigitcanlar, Madhav Nepal, Nguyen, K., & Fatih Dur. (2023, May 1). Machine Learning and

Remote Sensing Integration for Leveraging Urban Sustainability: A Review and Framework. https://doi.org/10.1016/j.scs.2023.104653

Quamar, M. M., Al-Ramadan, B., Khan, K., Shafiullah, M., & Ferik, S. E. (2023, October 1). Advancements and Applications of Drone-Integrated Geographic Information System Technology—A Review. Remote Sensing. https://doi.org/10.3390/rs15205039

Srivastava, S. (2023, September 21). AI, IoT, AR/VR revolutionizing technology for sustainability. Appinventiv. https://appinventiv.com/blog/technology-for-sustainability/

Taofeeq Ibn-Mohammed, Mustapha, K. B., Abdulkareem, M., A. Ucles Fuensanta, Pecunia, V., & Claire. (2023, October 5). Toward artificial intelligence and machine learning-enabled frameworks for improved predictions of lifecycle environmental impacts of functional materials and devices. MRS Communications; Springer Nature. https://doi.org/10.1557/s43579-023-00480-w

Wu, Y., & Tham, J. (2023, October 1). The impact of environmental regulation, Environment, Social and Government Performance, and technological innovation on enterprise resilience under a green recovery. Heliyon; Elsevier BV. https://doi.org/10.1016/j.heliyon.2023.e20278

public engagement - Search | Citizen Science: Theory and Practice. (2023). Citizen Science: Theory and Practice. https://theoryandpractice.citizenscienceassociation.org/search?q=public%20engagement&page=2

Conclusion

Throughout this book, we have explored the fundamentals of Environmental Impact Assessment (EIA), understanding its purpose in sustainable project planning. We began with the foundational concepts of EIA, outlining its critical role in identifying, predicting, and evaluating the potential environmental effects of proposed actions. Detailed aspects such as screening, scoping, baseline studies, impact prediction, mitigation measures, and monitoring were all dissected to provide readers with a comprehensive grasp of the procedural and practical elements involved.

The discussion included specific methodologies for conducting EIAs effectively, emphasizing the importance of robust data collection, thorough analysis, and stakeholder engagement. Each chapter delved into essential steps, including public consultation processes that ensure transparency and community involvement, which are crucial for achieving balanced decisions benefiting both development goals and environmental sustainability.

This book also discussed the legislative frameworks that govern EIAs globally, comparing various international standards and practices to highlight common principles and unique approaches. We examined case studies from different regions and sectors, illustrating how EIA has been successfully implemented in diverse contexts to mitigate negative impacts and enhance positive outcomes. These real-world examples underscored the practical application of theoretical knowledge, bridging the gap between academic learning and field practice.

Reiterating the significance of the EIA process is crucial in protecting the environment and promoting sustainable practices. The EIA is not merely a bureaucratic hurdle; it is a pivotal mechanism ensuring projects proceed with an awareness of their potential consequences—both positive and negative—on our planet. It serves as a scientific tool that empowers communities, guides responsible development, and underscores the urgency of safeguarding our natural resources. By systematically assessing environmental impacts, EIAs help prevent irreversible damage, advocate for mitigation strategies, and contribute to the balanced coexistence of development and conservation efforts.

The process sheds light on unforeseen environmental risks, enabling policymakers and project developers to make informed decisions that prioritize long-term ecological health over short-term gains. It reinforces the principle that sustainable development is achievable only when economic growth aligns with environmental stewardship. Furthermore, through public participation components, EIA fosters a democratic platform where diverse voices—including those of marginalized communities—are heard and considered in decision-making processes. This inclusivity strengthens the legitimacy and acceptance of projects, fostering communal trust and collaboration.

As you reflect on the principles and lessons detailed in this book, it is important to recognize the power your voice holds. Engagement with local communities and active participation in public consultations are essential steps toward ensuring every project considers both ecological and social impacts. Continued dialogue among stakeholders, including environmental professionals, policymakers, regulators, students, and academics, is vital for advancing EIA practices and adapting to emerging challenges.

Environmental issues are dynamic, evolving with scientific discoveries, technological advancements, and shifting societal values. As such, remaining engaged and informed is critical. Practitioners must continually update their knowledge and skills to address new environmental concerns effectively. Policymakers and regulators need to craft adaptive policies that reflect current realities while anticipating future trends. Academia plays a crucial role in researching and teaching innovative methods that push the boundaries of conventional EIA practices.

This collaborative effort ensures that EIA remains a living process—one that evolves and improves over time. It epitomizes the collective responsibility we share in preserving and enhancing our environment for future generations.

Together, let us commit to being stewards of our environment. By advocating for robust EIAs and adopting sustainable business practices, we can forge a future where development does not compromise our planet's integrity but rather enhances it for generations to come. This book equips you with the knowledge and tools needed to champion sustainable practices in your personal and professional lives. Now, it is up to each of us to take action.

Incorporate EIA principles in your daily decision-making processes, whether you are involved in large-scale projects or smaller initiatives within your community. Advocate for policies that enforce rigorous environmental assessments and support development projects demonstrating genuine commitment to sustainability. Educate others about the importance of EIAs and the broader implications of environmental stewardship.

Every action, no matter how small, contributes to the larger goal of achieving a sustainable and resilient future. Let us harness the knowledge gained from this book to inspire change and foster an enduring respect for our environment. Our collective efforts can drive significant improvements, ensuring that development projects worldwide adhere to high environmental standards.

By embracing the ethos of EIA, we honor our obligation to protect the earth's ecosystems, promote biodiversity, and secure a healthy environment for all living beings. Let us embolden our resolve to integrate sustainable practices into all facets of life, creating a legacy of responsible development deeply rooted in environmental consciousness.

In closing, the journey through this book should not end here. Let it be the beginning of a continuous quest for knowledge, understanding, and action in the realm of environmental impact assessment and sustainable development. Your commitment to these principles will shape a better world—one where economic progress and environmental resilience go hand in hand, leaving a thriving planet for the generations to come.

www.ingramcontent.com/pod-product-compliance
Ingram Content Group UK Ltd.
Pitfield, Milton Keynes, MK11 3LW, UK
UKHW062305290726
14090UKWH00018B/890

9 798895 440193